CW01376210

FLOWERING
indoor plants
Colin Rochford

£2.95

OORLOFF

FLOWERING
indoor plants

OORLOFF

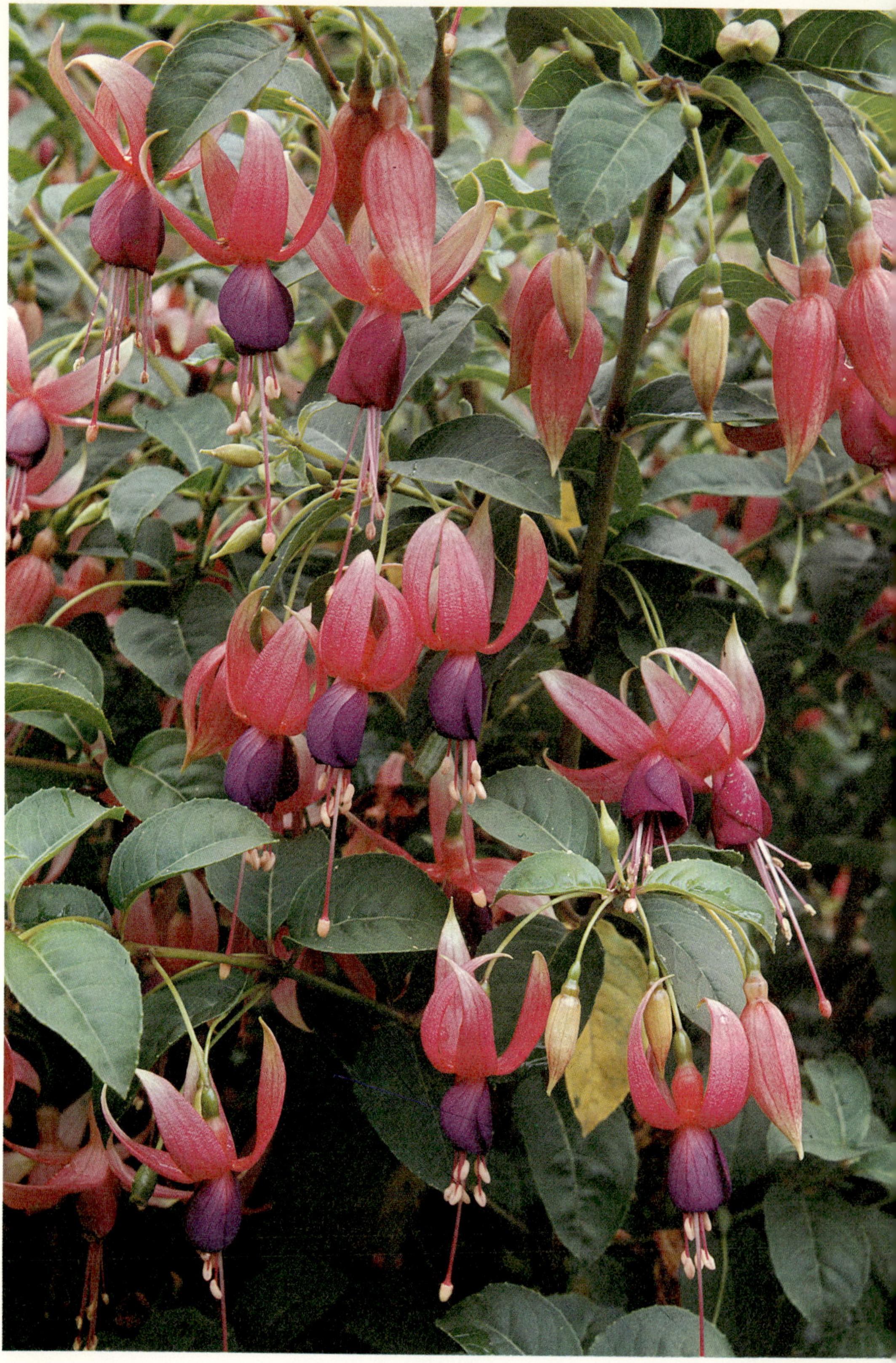

Contents

Introduction	11
Keeping Houseplants	12
Buying a plant	12
Planters and containers	12
Soil	12
Light	13
Temperature	14
Water	14
Humidity	15
Fertilizers and feeding	16
Pests and diseases	16
Propagation	17
General care	18
A–Z of Flowering Indoor Plants	20
Table of Flowering Seasons and Colour Types	90
Index of Common Names	92

This edition published by
Oorloff Books Limited
18 Priors Road, Windsor, Berkshire, England

© MCMLXXXIV Anthony Oorloff Publications Limited
ISBN 0 946883 02 5

Produced by
The Man Himself - England

© Spanish Edition. Editorial Tiempo Libre, S.A.
D.L.: M-35.932-1984

Printed in Spain.
By Alvi Industrias Gráficas, S.A. Madrid.

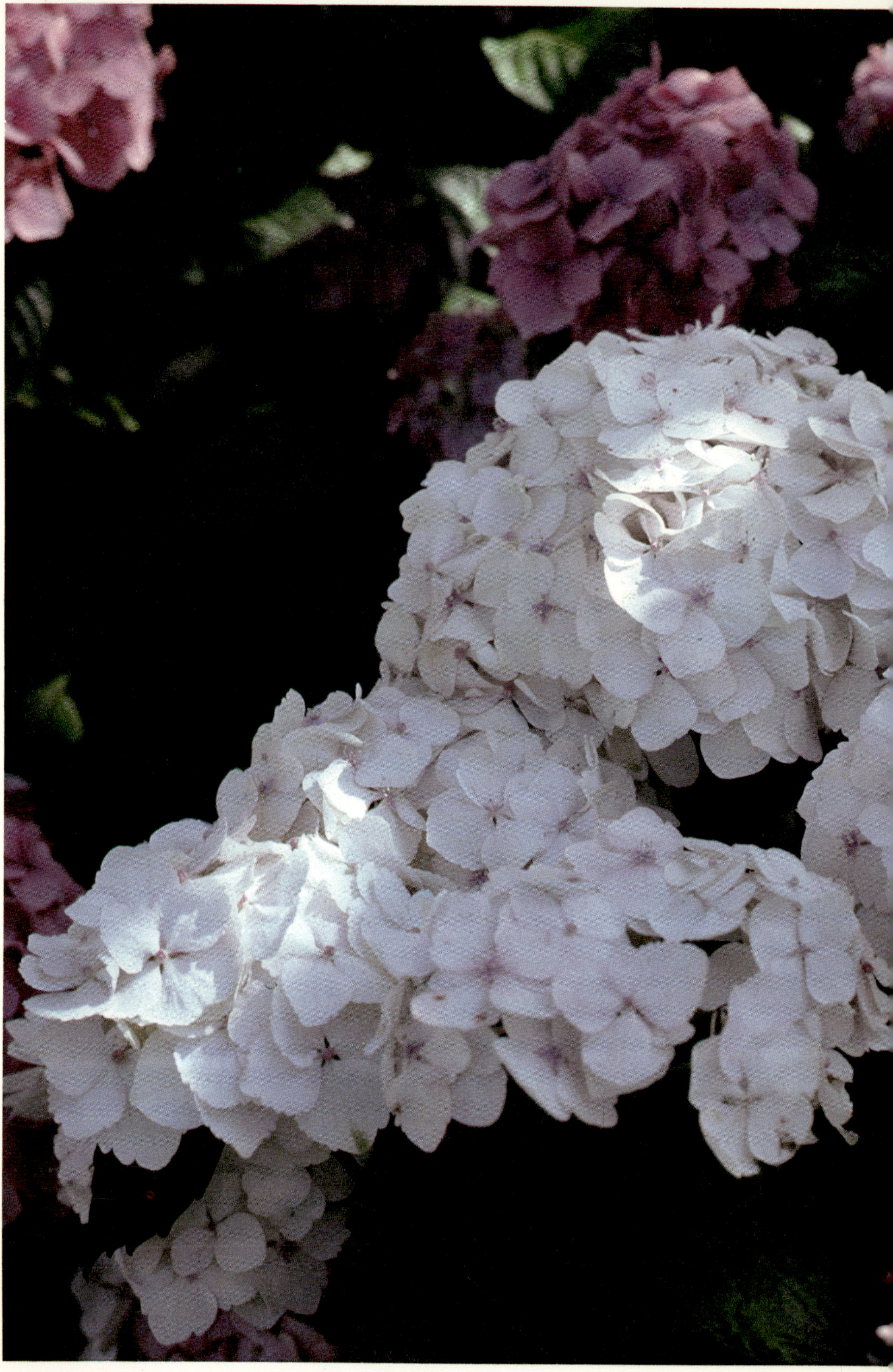

Introduction

Flowers have been around for some time. With improvements to the environment and communications, greater varieties of species are now easily available – from the more common forms to the exotic. Be they cut or in pots, flowers are a sight to behold. Therefore, we tend to look for ways to prolong the joy they give.

Plants have certain requirements to trigger-off their flowering cycles and by meeting these budding can be induced, turning what in essence is a reproductive function into a pleasurable advantage to both plant and grower.

Keeping Houseplants

Flowers are eye-catching and, for most of us, an introduction to the houseplant is through a dazzling display in a shop or shop-window. The appeal is too strong to overcome but a hasty purchase can later prove to be disastrous.

Understanding a plant's requirements is essential to obtain a reasonable amount of success.

Buying a plant

Before buying a plant, consider where you intend to place it and whether this position is suitable for its growth habits. Some plants revel in bright sunlight, whilst others prefer slightly shaded or cool situations. If you can satisfy these preferences then proceed with your choice.

The chances are that the houseplants in the shop were raised under glass in controlled conditions. Ensure that they have been 'hardened off'. This enables a plant to withstand the change in conditions from 'factory' to household.

Check each specimen for damaged leaves or flowers and pests. The plant should be sturdy, have darkish leaves and be free of damage and pests. If the weather is cold outside, ask the seller to wrap up the whole plant. Some shops do this anyway and, in most cases, these establishments usually prove to be reputable.

When taken indoors, treat tenderly for a week, keeping out of draughts and direct sunlight. Avoid the temptation to over-water. After this probationary period, it will be possible to place the plant in its permanent position and treat it as shown in the following pages.

Planters and containers

The majority of plants come in standard-shaped plastic pots. Nowadays, plants form part of the decor and there is a tendency to use decorative containers and planters with matching saucers to disguise the rather drab-looking plastic pot. Always use one which has a drainage hole as this allows any excess water to collect in the saucer. This surplus should be removed, otherwise plant problems will occur.

Soil

The two compost mixtures commonly used have either peat or loam as a base. Young plants and most flowering species prefer a peat-based composition. Although these mixtures can be made up, it is advisable to use one of the proprietary brands available. Do not use garden soil unless it has been sterilized.

Keeping Houseplants

Light

Lighting levels play an important part in the development of plants, especially in the formation of flowering buds. These levels can be assessed as follows:

Very high
A sunny window-sill receiving sunlight, i.e. a window facing south.

High
A position within 1.2 m (4 ft) of a large window facing south, east or west.

Medium
A well-lit area 1.2-2.4 m (4-8 ft) away from a large window.

Low
A position in excess of 2.4 m (8 ft) away from a window with no direct light, but not in total darkness.

If artificial lighting is used, keep the light source away from the plants or the heat emitted will cause leaf scorch. A minimum distance of at least 1.2 m (4 ft) should be maintained and the light switched on for periods of up to eight hours daily.

Some common light problems

Symptom	Probable cause	Remedy
Yellowing of leaves or leaves drop	Too much light	Reposition plant to a less sunny spot
Yellow to brown patches on leaves	Sunburn	Provide a sunscreen or move out of direct sunlight
Weak, soft and thin growth and leaves drop	Lack of light	Increase light intensity, prolong exposure to light
Wilting leaves	Too much light	Move plant further away from the light source

Keeping Houseplants

Temperature

Most indoor plants are native to warm climates. The warmer it is, the brighter and more exotic the flower. Some plants require a drop in the night temperature for their development. However, the plants now available can tolerate temperature ranges in the day of 18-26°C (65-79°F) and in the night of 10-21°C (50-70°F).

Some common temperature problems

Symptom	Probable cause	Remedy
Bud drop, small new leaves; weak, soft and thin growth	Temperature high at night	Reduced temperature by 5°C (10°F)
Yellowing of leaves or spotting	Temperature too low	Increase room temperature; give fresh air

Water

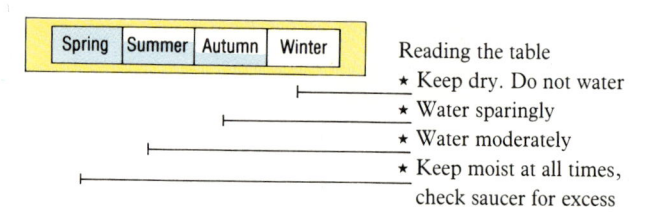

Spring	Summer	Autumn	Winter

Reading the table
★ Keep dry. Do not water
★ Water sparingly
★ Water moderately
★ Keep moist at all times, check saucer for excess

Understanding the water requirement of a plant is one of the essentials for success. Many plants are lost due to over/under-watering. Get used to inspecting the plant weekly or preferably daily (especially in the warmer months). Some plants need to rest in the cooler months and watering should be avoided at these times. Never let water collect in the saucer.

Some common water problems

Symptom	Probable cause	Remedy
Brown leaf tips or margins	Lack of water	Check soil for dryness; provide moisture
Leaves drop or curl	Draughts	Check position and change situation
Wilting leaves	Root damage	Check for excessive root growth and repot if necessary

Keeping Houseplants

Humidity

Water is lost through minute pores in the leaves. If not replaced, this could lead to drying out. The drier the conditions, the greater the loss. This can be overcome by misting the leaves or, in certain cases, providing a micro-climate around the plant.

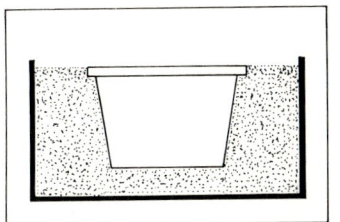

Choose a container which is taller than the existing pot. Fill the bottom with moist peat and place the plant in the vessel. Continue packing the sides with the peat until the top is reached. Firm down and level off. Keep the peat moist at all times. Use this method for taller specimens.

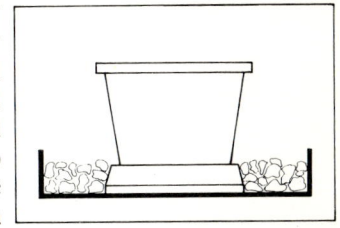

Use a larger plant saucer than normal and half-fill with small stones. To support the plant, place a small saucer on the base in an inverted position. Fill with water to a level below the bottom of the plant pot. Replace any water lost through evaporation. Use this method for small specimens.

Various plants have different humidity levels, and these can be summed up as follows:

Very high
Plants would benefit if an artificial climate was created.

High
Mist spray very frequently.

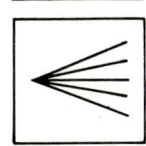

Medium
Mist spray at least twice a month, especially in the growing period.

Low
Little or no misting required.

15

Keeping Houseplants

Fertilizers and feeding

Every 14 days			
Spring	Summer	Autumn	Winter

Reading the table
★ The table indicates frequency and the optimum period when feeding should be carried out. Follow the instructions very carefully.

Organic or inorganic fertilizers can be used to replace the plants' constant usage of the elements. Always use the proprietary brands and follow the instructions carefully.

Pests and Diseases

Many flowering indoor plants are lost through faulty culture. Here are some of the common symptoms and remedies.

Aphids (pest)
Tiny green, orange, black or grey insects which feed by sucking out the sap of plants. Found in groups under leaves or on stems. They excrete a sticky liquid called 'honeydew' which attracts a black fungus (sooty mould). Treat the infected area by wiping with a weak soapy solution. Rinse with fresh, clean water. Malathion applied fortnightly is equally as effective.

Botrytis (disease)
A grey cotton-wool-like fungus which attacks the whole plant. Can be prevented by removing all dead or dying leaves as they become evident. This fungus often lives on the rotting, discarded parts, and if conditions become either too cool, less airy or too humid, the fungus will infect the plant. Cut away infested portions and spray the area with a fungicide weekly for a month.

Damping-off (disease)
A disease usually associated with cuttings and seedlings. The stems or roots begin to rot away with resulting wilting. The result of the use of unsterilized potting mixtures. Use a solution of water and Benomyl once a week for at least three weeks and move plants to an airy, cooler location.

Mealy bugs (pest)
Small white insects resembling woodlice. The young insects are developed in white, fluffy webs. Tendency to inhibit leaf joints and central leaf margins where they feed. Can be hard to eradicate, especially if unnoticed and allowed to settle. Treat by dabbing the insects with methylated spirits. Use a fine brush and douse the offending pest. Malathion, used as instructed, can also prove effective.

Keeping Houseplants

Mildew (*disease*)
This disease falls into two main categories: a powdery type which looks like a fine white dust, and a downy type which has a fluffy, cotton-wool appearance. Usually caused by bad conditions. Dust with sulphur twice a week or spray with a fungicide.

Red spider mites (*pest*)
Extremely tiny sap-sucking insects which are almost indiscernible to the naked eye. They congregate on the undersides of leaves which then develop yellow spots and later drop off. Look for a fine web between the leaves and stems, as this is a sure indication of their presence. Hot, dry conditions produce rapid infestation. If found, spray weekly with Malathion.

Scale insects (*pest*)
Blister-shaped tiny brown insects which fasten themselves to the undersides of leaves and stems. If the plant is badly infested, they will also be found on upper leaf surfaces. Best trated by either removing with a fingernail or by spraying once a week for a month with Malathion. Should a leaf be severely infested, it is always better to remove it completely.

Thrips (*pest*)
Tiny black insects which attack the buds, flowers and leaves. They cause loss of flower buds and a silvery discoloration of the leaves. Damaged leaves or flowers are best cut off. Malathion sprayed once a week for two weeks is effective.

White flies (*pest*)
It is the young, green larvae of these pale insects which are the real culprits. They attach themselves to the undersides of the leaves where they suck out the sap whilst excreting the tell-tale glistening honeydew which acts as a medium for a black fungus. Infested leaves turn yellow and fall. Adults flit from plant to plant laying eggs. Spray with Malathion twice a week for a month.

Propagation

Whilst the majority of flowering plants are reproduced by seed or cuttings, there are also various other methods of propagation.

Division
With division, the plant is taken out of the pot. The clump of root is then broken apart into two or three portions as required. Care and attention is needed. Repot the divided plant sections. Some roots may need to be cut with a sharp knife as they form dense root growths. In such cases, any open wound must be dusted with a fungicide to prevent disease.

Leaf cuttings
In this method, the leaves are cut into 10 cm (4 in) sections, potted into small pots and put into a propagation frame. Another way is to make small incisions through the main vein in a leaf and place it flat on to a rooting compost. New plants will develop from each cut made.

Keeping Houseplants

Leaf-bud cuttings
Cut off a section of stem with one or more leaf nodes attached. Pot the severed piece in a rooting medium. Roots and new growth will appear within a few weeks.

Offsets (self-propagation)
Some houseplants develop long trailing stems at the end of which young plantlets are produced. These plantlets should be pegged down in compost and, once the young shoots have established themselves, the long stem can be cut from the parent. With bromeliads, the young side shoots produced when the plant is in flower should only be potted after the mature plant has died.

Seed-sowing
Select a seed tray and first place gravel or broken pieces of a clay pot at the bottom (this will give good drainage). Top this with a layer of seeding compost until the soil level is about 3 cm (1 in) from the top. Press the compost down firmly to ensure an even depth. Water the compost till it is well drenched by using a fine spray. Sow the seeds by either scattering them or placing them in rows, depending on the size of seed used. Once sown, place a small layer of compost over the seeds and cover the tray with a sheet of paper topped by a sheet of glass. Seedlings are sensitive to harsh light when they begin to sprout. Position the seed tray in a warm but ventilated place. Damping-off can be avoided by dusting over the pane of glass each day. Once a second pair of leaves has appeared, the seedlings are ready to be potted-on.

Stem/terminal cuttings
Choose a strong, firm stem. Cut off about 7 cm (3 in) just below a leaf joint. Remove the lower leaves, dust with a rooting compound and place in the rooting medium which should contain at least 30% sharp sand. Water the cuttings and place in a propagator. Spray from above and ventilate daily. After a month, roots will have formed, and the young plants can be potted. An alternative method is to place the cuttings in water as the developing roots can be seen more easily. Then treat as detailed above.

General care

Plants will benefit from regular pinching of tips, pruning, the removal of dead flower-heads and occasional repotting.

Repotting or potting-on
This is simply the transfer of an existing plant to a larger pot. Many healthy plants soon outgrow their existing containers. Repotting should only be carried out when it is required. Some general signs to look out for are:
(1) roots appear out of the drainage hole;
(2) growth becomes slow despite proper feeding;
(3) soil dries out quickly and watering is needed frequently.
If you are not certain, try the following test. Do not water the plant for a day and allow the soil to become slightly dry. Place the palm of your hand across the top of the pot (fingers

spread out to enable the stem to protrude) and turn the pot plant upside down. Gently tap the side of the pot. The soil ball should separate from the pot. If it is pot-bound, there will be masses of roots and hardly any soil, in which case repotting is necessary. Otherwise, return the plant and soil ball to the pot.

How to repot

The ideal time to repot is in the late spring, as this will give the roots ample time to establish themselves before the onset of autumn. Pick a pot which is slightly larger than the existing one and ensure that it is clean. Scrub if necessary.

The drainage hole should then be covered up with pieces of old pots or broken bricks (depending on the size of the plant).

Cover these crocks with peat and top up with a layer of the potting medium to be used.

Remove the plant from its old pot as described earlier. Trim back any excessive root growth and carefully detach the old pieces of broken crocks or stones. Having done so, place the plant into the new pot. Fill in the sides with the potting medium until you reach the top soil level of the plant. Firm gently as you go along. Water and place the new pot and plant in a shady area for about six days. To avoid wilting, spray the leaves daily. The plant is now ready to be replaced in its original position and cared for accordingly.

A-Z of Flowering Indoor Plants

In the following pages, in alphabetical order, a selection of popular flowering houseplants is given. Each species is illustrated, together with a general description and recommendations for growing, and a chart provides information for easy reference.

Aphelandra squarrosa 'Dania'

Brazil

Zebra plant

Acanthaceae

A striking plant in or out of flower. The yellow/orange-tipped bracts against the creamy-veined, dark green leaves make this a popular choice. Keep away from draughts or loss of the lower leaves will result. This plant produces masses of roots and should therefore be potted-on every spring. After flowering, remove the dead bract and cut back to the next leaf axil. New growth when formed can be potted in a peat-based medium and should root within five to eight weeks.

Colour Types
Predominant single colour
　　Blue
　　Orange
　　Pink
　　Purple/Mauve
　　Red
　　White
　　Yellow ✻
　　Multicoloured

Avoid direct sunlight and draughts

| Spring | Summer | Autumn | Winter |

Loam-based

Aphids, mealy bug, red spider

16–20°F (61–68°F)

Every 14 days

| Spring | Summer | **Autumn** | Winter |

Autumn
Leaf-bud cuttings

Flower colour types

20

A-Z of Flowering Indoor Plants

Azalea

Japanese rose

Japan —— Country/Place of origin

Ericaceae ——— Family

The azalea is a beautiful evergreen shrub with small ovoid leaves and flowers that range in colour from red, pink and mauve to white. When grown indoors, choose a well-lit spot out of direct sunlight. In early spring, after it has flowered, new shoots will develop. These should be cut out but any shoots forming after late spring should be left. The plant may then be placed outdoors for the summer. By early autumn buds will have formed and the plant should be brought inside and placed at a window (if positioned immediately in a warm room bud-drop will occur, a window is usually cooler). Spray regularly to maintain humidity. When watering, use softened water or collected rainwater, and water generously when the plant is flowering, but give less water during the resting period. Only feed the plant in summer. When growing cuttings, a greenhouse is always necessary.

Temperature:
Range

Humidity:
Very high
High
Medium
Low

Fertilizer:
Frequency
Seasons

Flowering period
Propagation

Colour Types
  Predominant single colour
    Blue
    Orange
    Pink ✯
    Purple/Mauve ✯
    Red ✯
    White ✯
    Yellow
  Multicoloured

Avoid direct sunlight

Spring | Summer | Autumn | Winter

Peat-based

Red spider

5–15°C (41–59°F)

Every 14 days
Spring | Summer | Autumn | Winter

Mid winter to mid spring
Cuttings

✯ Recommended ideal conditions

21

Acalypha hispida

New Guinea

Red hot cat's tail

Euphorbiaceae

The feature of this plant is its long red bracts. Keep in a bright position but out of direct sunlight. The soil should be kept moist and a high degree of humidity maintained. The plants are very difficult to keep throughout the winter unless you have adequate heating conditions. Take terminal cuttings in the spring. Cut back at this time to rejuvenate older plants.

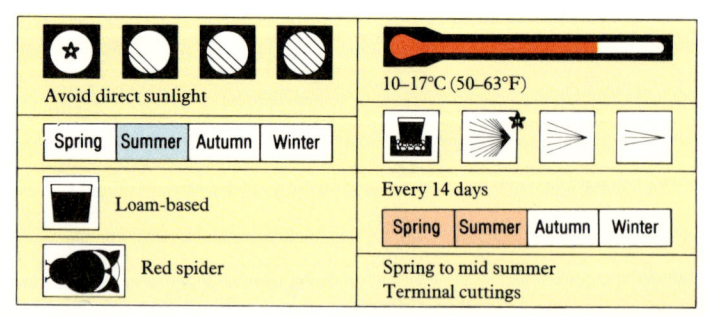

Colour Types
    Predominant single colour
        Blue
        Orange
        Pink
        Purple/Mauve
        Red ✱
        White
        Yellow
    Multicoloured

Avoid direct sunlight

| Spring | Summer | Autumn | Winter |

Loam-based

Red spider

10–17°C (50–63°F)

Every 14 days

| Spring | Summer | Autumn | Winter |

Spring to mid summer
Terminal cuttings

Aechmea fasciata

South America

Urn plant

Bromeliaceae

This is a very popular bromeliad with large, tough, silver-grey/green marbled leaves and a pink bract with tiny blue flowers which later fade to pink. The flowerhead can last for months. When it is in flower the foliage is in the process of decay. Tiny offsets will form at this time and should only be removed to be potted-on when the leaves of the parent plant show signs of rotting. These young shoots should flower within two years. However, to induce flowering, enclose the plant with a ripening apple in an airtight plastic bag for four to five days. If there is no response within two months, repeat the process.

Requires a well-lit position. Keep moist by filling the funnel of the plant with water and allowing the excess to spill over into the soil.

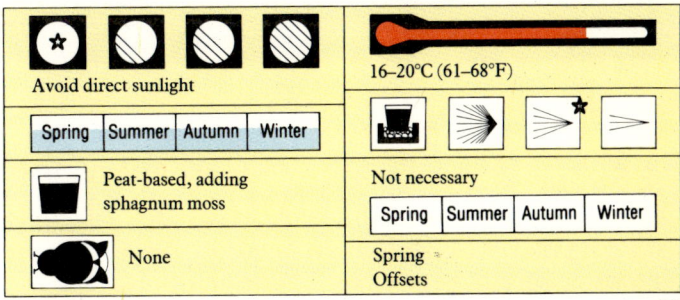

Colour Types
Predominant single colour
Blue
Orange
Pink ✪
Purple/Mauve
Red
White
Yellow
Multicoloured

Avoid direct sunlight

| Spring | Summer | Autumn | Winter |

Peat-based, adding sphagnum moss

None

16–20°C (61–68°F)

Not necessary

| Spring | Summer | Autumn | Winter |

Spring
Offsets

23

Aeschynanthus

Basket vine

Java

Gesneriaceae

A. splendidus

This attractive plant sports shiny orange/red-tipped flowers from calyces. Requires a high temperature in the growing period. Use tepid water and maintain a high degree of humidity. Cuttings need extra bottom heat. *A. radicans* (lipstick vine) is a trailing plant with clusters of brilliant scarlet tubular flowers which blossom in the summer. Propagation is by division or terminal cuttings.

Aeschynanthus

Lipstick vine

Java

Gesneriaceae

A. radicans

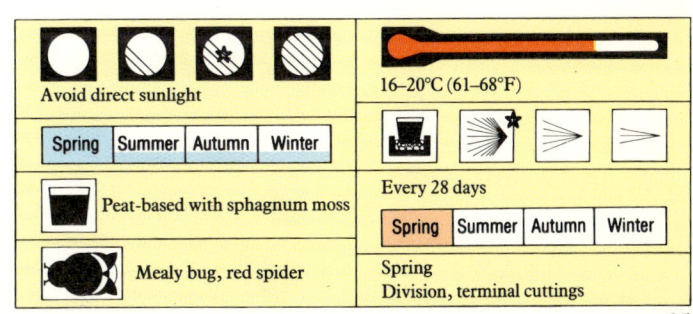

Colour Types
Predominant single colour
Blue
Orange
Pink
Purple/Mauve
Red ✭
White
Yellow
Multicoloured

Avoid direct sunlight

| Spring | Summer | Autumn | Winter |

Peat-based with sphagnum moss

Mealy bug, red spider

16–20°C (61–68°F)

Every 28 days

| Spring | Summer | Autumn | Winter |

Spring
Division, terminal cuttings

25

Allamanda cathartica

Brazil

Golden trumpet

Apocynaceae

A climbing plant with spear-shaped foliage and extremely striking golden yellow, funnel-shaped flowers. Should be kept in bright light, moist in the growing period, dry in winter. Cut back by at least a third after the flowers have died. Stem cuttings can be taken in mid-spring from mature growth.

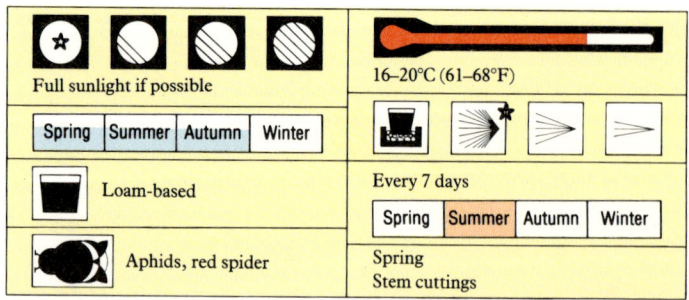

Colour Types
    Predominant single colour
        Blue
        Orange
        Pink
        Purple/Mauve
        Red
        White
        Yellow ✶
    Multicoloured

Full sunlight if possible

Spring | Summer | Autumn | Winter

Loam-based

Aphids, red spider

16–20°C (61–68°F)

Every 7 days

Spring | Summer | Autumn | Winter

Spring
Stem cuttings

Anthurium scherzerianum   Costa Rica

Flamingo plant   Araceae

The long, narrow, lance-shaped leaves, coupled with its curled flower cluster attached to flat red bracts, make this a houseplant whose popularity is increasing. The soil should contain at least a third of sphagnum moss if the plants are to thrive. Keep the soil moist but water less in the winter. Plants enjoy a high degree of humidity, so mist regularly.

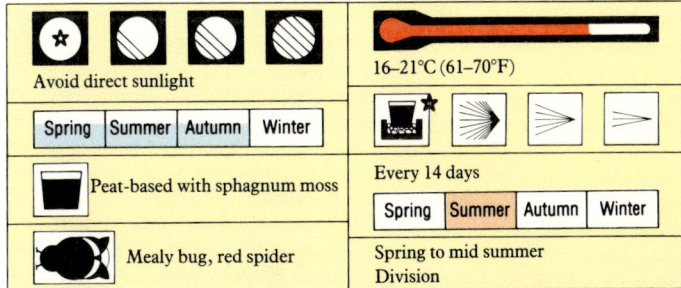

Colour Types
    Predominant single colour
        Blue
        Orange
        Pink
        Purple/Mauve
        Red ✱
        White
        Yellow
    Multicoloured

Avoid direct sunlight

| Spring | Summer | Autumn | Winter |

Peat-based with sphagnum moss

Mealy bug, red spider

16–21°C (61–70°F)

Every 14 days

| Spring | Summer | Autumn | Winter |

Spring to mid summer
Division

Aphelandra squarrosa 'Dania'

Zebra plant

Brazil

Acanthaceae

A striking plant in or out of flower. The yellow/orange-tipped bracts against the creamy-veined, dark green leaves make this a popular choice. Keep away from draughts or loss of the lower leaves will result. This plant produces masses of roots and should therefore be potted-on every spring. After flowering, remove the dead bract and cut back to the next leaf axil. New growth when formed can be potted in a peat-based medium and should root within five to eight weeks.

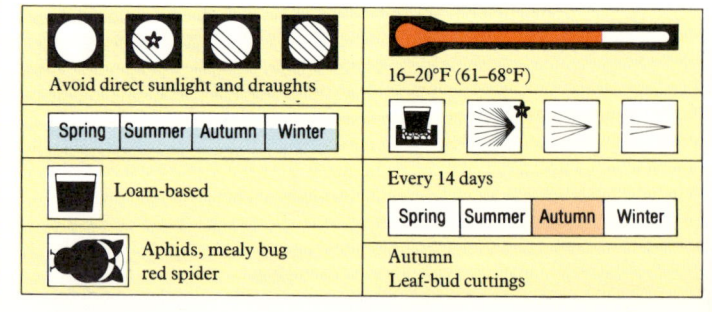

Colour Types
  Predominant single colour
    Blue
    Orange
    Pink
    Purple/Mauve
    Red
    White
    Yellow ✶
  Multicoloured

Avoid direct sunlight and draughts

| Spring | Summer | Autumn | Winter |

Loam-based

Aphids, mealy bug, red spider

16–20°F (61–68°F)

Every 14 days

| Spring | Summer | Autumn | Winter |

Autumn
Leaf-bud cuttings

Azalea

Japanese rose

Japan

Ericaceae

The azalea is a beautiful evergreen shrub with small ovoid leaves and flowers that range in colour from red, pink and mauve to white. When grown indoors, choose a well-lit spot out of direct sunlight. In early spring, after it has flowered, new shoots will develop. These should be cut out but any shoots forming after late spring should be left. The plant may then be placed outdoors for the summer. By early autumn buds will have formed and the plant should be brought inside and placed at a window (if positioned immediately in a warm room bud-drop will occur, a window is usually cooler). Spray regularly to maintain humidity. When watering, use softened water or collected rainwater, and water generously when the plant is flowering, but give less water during the resting period. Only feed the plant in summer. When growing cuttings, a greenhouse is always necessary.

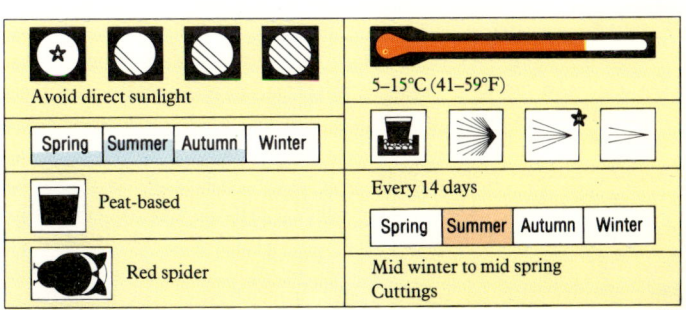

Colour Types
  Predominant single colour
    Blue
    Orange
    Pink ✿
    Purple/Mauve ✿
    Red ✿
    White ✿
    Yellow
  Multicoloured

Avoid direct sunlight

| Spring | Summer | Autumn | Winter |

Peat-based

Red spider

5–15°C (41–59°F)

Every 14 days

| Spring | Summer | Autumn | Winter |

Mid winter to mid spring
Cuttings

Begonia

South America

Paper petals

Begoniaceae

A species that is one of the most popular of flowering houseplants. The colours available are white, pink, red, yellow and orange. The flower sizes and shapes are varied. Tuberous begonias include small- or large-flowered and trailing forms. The tubers are kept dry in winter and gradually brought to life in moist peat in early spring, and then potted. After the plants have stopped growing, water is withheld and by late summer the stems are removed. The tubers are then taken out and left to dry, then stored in dry peat fibre. Another group of begonias are the winter-flowering types. They are called 'winter-flowering' because it used to be considered odd that these hybrids flowered in winter. This group also includes both small- and large-flowered forms. The flowers are usually single, although semi-double or double types are available. These hybrids have a very long flowering season and come in a variety of colours. Blooms are encouraged by artificially prolonging daylight using 60 watt bulbs suspended above the plant at about 50-80 cm (20-30 in) and leaving them on for about sixteen hours a day. As a result of this the plants will flower in winter. When growing begonias, a well-lit position out of direct sunlight is desirable. Always keep the soil moist but allow to dry between waterings and feed fortnightly, but withhold water and do not feed during the resting period. Should plants become ragged, cutting back will keep them in shape.

Begonia

South America

Paper petals

Begoniaceae

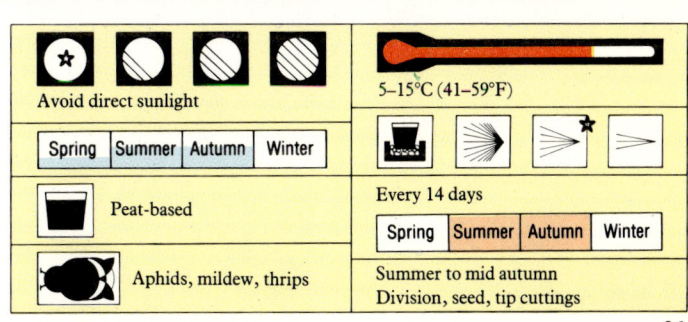

Colour Types
- Predominant single colour
  - Blue
  - Orange ✱
  - Pink ✱
  - Purple/Mauve
  - Red ✱
  - White ✱
  - Yellow ✱
- Multicoloured

Avoid direct sunlight

| Spring | Summer | Autumn | Winter |

Peat-based

Aphids, mildew, thrips

5–15°C (41–59°F)

Every 14 days

| Spring | Summer | Autumn | Winter |

Summer to mid autumn
Division, seed, tip cuttings

Beloperone guttata

Mexico

Shrimp plant

Acanthaceae

Produces pale pink bracts whose overall shape gives the plant its common name. The short-lived white flowers are hidden by the bracts, although the blooms last well into autumn. Place in a sunny spot to encourage full colour in the bracts. Keep the soil moist throughout the year and avoid extremes of temperature. These plants do not really need feeding, but a fortnightly feed does them no harm. A plant that stands up well to the dry air conditions caused by central heating. Can be placed outdoors during the summer. Pruning the top growth in early summer will produce bushy plants. Any straggly stems can also be cut back at this time.

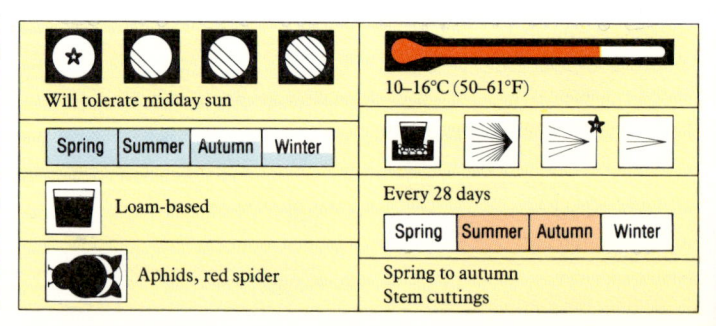

Colour Types
Predominant single colour
Blue
Orange
Pink ✱
Purple/Mauve
Red
White
Yellow
Multicoloured

Will tolerate midday sun

| Spring | Summer | Autumn | Winter |

Loam-based

Aphids, red spider

10–16°C (50–61°F)

Every 28 days

| Spring | Summer | Autumn | Winter |

Spring to autumn
Stem cuttings

Bougainvillea

Brazil

Paper flower

Nyctaginaceae

An attractive climbing houseplant with light to dark green leaves and colourful floral bracts which encircle completely the inconspicuous flowers. During the summer months the plant should be kept outdoors in a warm, sunny spot to encourage flowering. During the growing season the plant requires plenty of water and feeding fortnightly, but water sparingly and do not feed in winter when the plant should be brought inside. Cuttings should be taken in spring. To keep the plant healthy, spray the wood. Do not worry if the leaves fall, as this is usual. *Bougainvillea spectabilis* grows very tall and has a thorny stem.

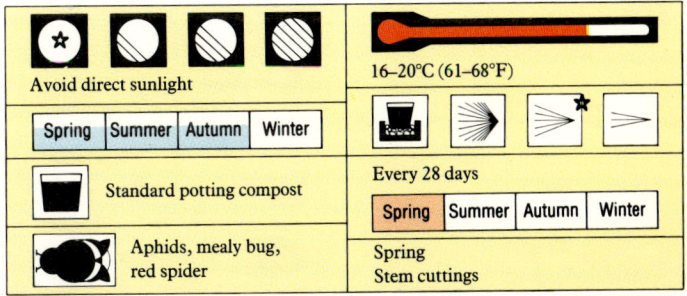

Colour Types
- Predominant single colour
  - Blue
  - Orange
  - Pink
  - Purple/Mauve ✯
  - Red
  - White
  - Yellow
- Multicoloured

Avoid direct sunlight

| Spring | Summer | Autumn | Winter |

Standard potting compost

Aphids, mealy bug, red spider

16–20°C (61–68°F)

Every 28 days

| Spring | Summer | Autumn | Winter |

Spring
Stem cuttings

33

Cactaceae

South America

Cactus   Cactaceae

Cacti make excellent indoor plants as most are able to survive short periods of neglect better than other species. There are about 2,000 species. The majority of cactus plants have spines and the characteristic areole from which emerges flowers, bristles, hairs, shoots or wool. Keep in a sunlit position and maintain a high temperature, although the plants will benefit from a slightly cooler environment in the winter. A common loss of plants is due to over-watering. The soil should be kept as dry as possible with infrequent waterings.

Conophytum pearsonii

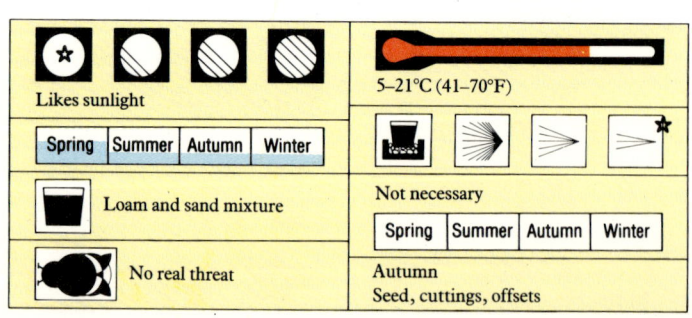

Colour Types
Predominant single colour
Blue
Orange
Pink
Purple/Mauve ✻
Red
White
Yellow
Multicoloured

Likes sunlight

Spring	Summer	Autumn	Winter

Loam and sand mixture

No real threat

5–21°C (41–70°F)

Not necessary

Spring	Summer	Autumn	Winter

Autumn
Seed, cuttings, offsets

Cactaceae

South America

Cactus Cactaceae

Mammillaria zeilmanniana

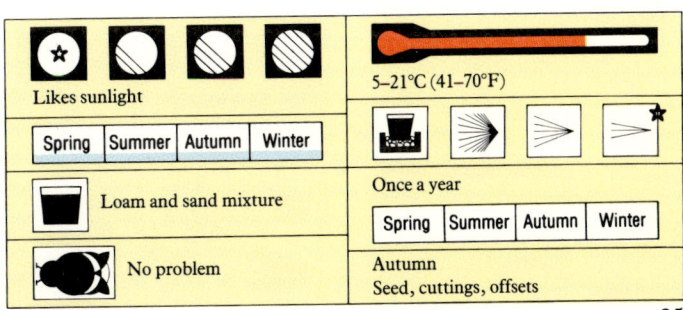

Colour Types
Predominant single colour
Blue
Orange
Pink
Purple/Mauve
Red ✱
White
Yellow
Multicoloured

Likes sunlight

| Spring | Summer | Autumn | Winter |

Loam and sand mixture

No problem

5–21°C (41–70°F)

Once a year

| Spring | Summer | Autumn | Winter |

Autumn
Seed, cuttings, offsets

Cactaceae

South America

Cactus   Cactaceae

Opuntia macrohiza

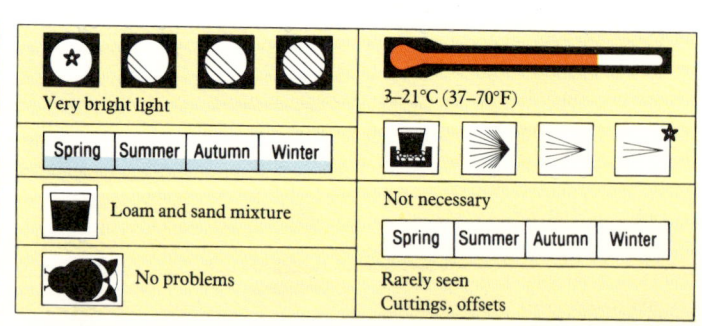

Colour Types
- Predominant single colour
  - Blue
  - Orange
  - Pink
  - Purple/Mauve
  - Red
  - White
  - Yellow ✯
- Multicoloured

Very bright light

Spring | Summer | Autumn | Winter

Loam and sand mixture

No problems

3–21°C (37–70°F)

Not necessary

Spring | Summer | Autumn | Winter

Rarely seen
Cuttings, offsets

Cactaceae

South America

Cactus   Cactaceae

Rubutia senilis

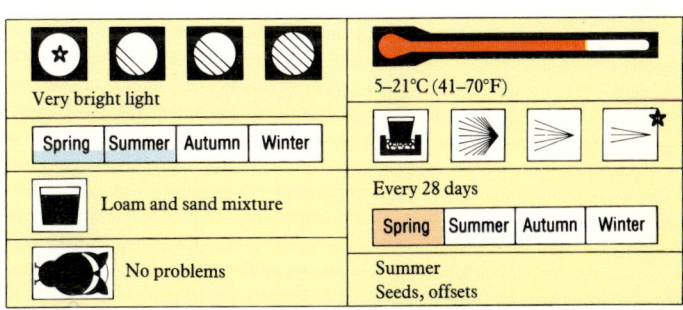

Colour Types
Predominant single colour
Blue
Orange
Pink
Purple/Mauve
Red ✿
White
Yellow
Multicoloured

Very bright light

Spring | Summer | Autumn | Winter

Loam and sand mixture

No problems

5–21°C (41–70°F)

Every 28 days

Spring | Summer | Autumn | Winter

Summer
Seeds, offsets

37

Calceolaria

Hybrid

Lady's slipper

Scrofulariaceae

This plant has large, hairy leaves with contrasting spotted flowers. The flowers have upper and lower lips, the latter being grossly enlarged. Once the flowers have faded the plant is best disposed of, although keeping the plant cool and feeding with liquid fertilizer will help to give a longer flowering time. Avoid positioning the plant in a draught as this will attract aphids, especially if the conditions are too warm. Seeds are best sown in spring or summer. Calceolaria is available in a variety of colours including yellow and red.

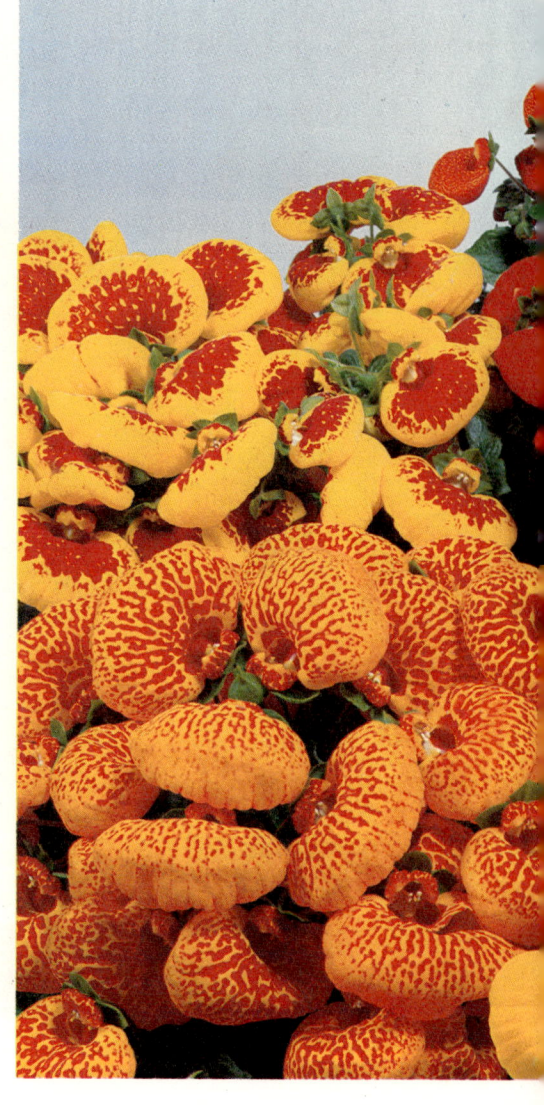

Calceolaria

Hybrid

Lady's slipper

Scrofulariaceae

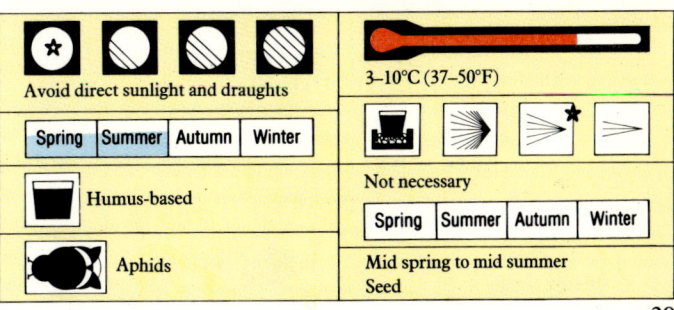

Colour Types
- Predominant single colour
  - Blue
  - Orange ✱
  - Pink
  - Purple/Mauve
  - Red ✱
  - White
  - Yellow ✱
- Multicoloured

Avoid direct sunlight and draughts

| Spring | Summer | Autumn | Winter |

Humus-based

Aphids

3–10°C (37–50°F)

Not necessary

| Spring | Summer | Autumn | Winter |

Mid spring to mid summer
Seed

39

Camellia japonica

Tea plant

Far East

Theaceae

A very attractive evergreen plant with rosettes of fine flowers and oval, dark green leaves whose edges are serrated. Colouring ranges from white to red. Camellias grow best in a bright position. Try to keep temperatures fairly constant as these are sensitive plants. Keep the soil moist using tepid water, never allow it to dry out or bud-drop will occur. Spray the plants frequently to maintain a humid environment. Feed during the spring only, when the plants are most active. Tip cuttings should be taken at this time. After the plant has flowered, it can be placed outside in a cool but frost-free position.

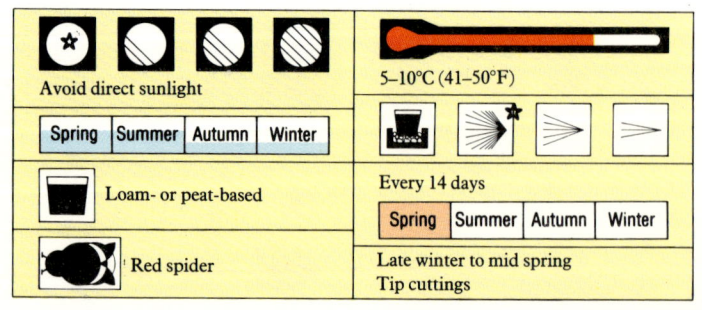

Colour Types
  Predominant single colour
    Blue
    Orange
    Pink ✿
    Purple/Mauve
    Red ✿
    White ✿
    Yellow
  Multicoloured

Avoid direct sunlight

| Spring | Summer | Autumn | Winter |

Loam- or peat-based

Red spider

5–10°C (41–50°F)

Every 14 days

| Spring | Summer | Autumn | Winter |

Late winter to mid spring
Tip cuttings

Canna indica

South America

Canna Cannaceae

The flowers are tuft-like, produced at the top of stems. The red and yellow varieties are most popular. The leaves are spear-shaped and sessile (without stalks). Grow canna in bright and sunny spots. These plants tend to lose a lot of moisture through evaporation, so in fairly hot weather give plenty of water. Repotting annually in spring is beneficial. The offsets produced are best left to the most well-informed of plant-growers.

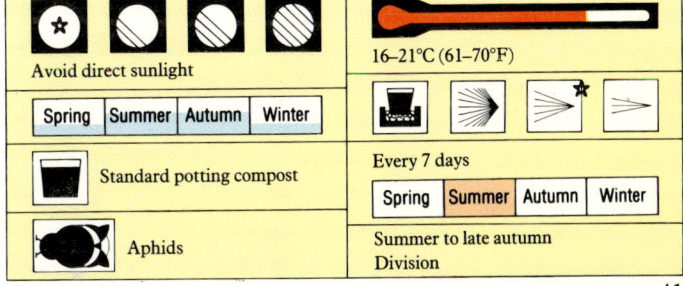

Colour Types
   Predominant single colour
      Blue
      Orange
      Pink
      Purple/Mauve
      Red ✿
      White
      Yellow ✿
   Multicoloured

Avoid direct sunlight

| Spring | Summer | Autumn | Winter |

Standard potting compost

Aphids

16–21°C (61–70°F)

Every 7 days

| Spring | Summer | Autumn | Winter |

Summer to late autumn
Division

Catharanthus roseus

Madagascar

Madagascar periwinkle

Apocinaceae

Dark green, oval, white-veined leaves and the rosy-centred white flowers make this a fine addition to the home. These plants need a bright and sunny spot if they are to produce good colour. Keep the soil damp, and ensure that the plant does not dry up. During the summer months feed fortnightly but in winter withhold feed. Maintain a humid environment. Tip cuttings should be taken in spring after the plant has been allowed to winter at 10-20°C (50-68°F).

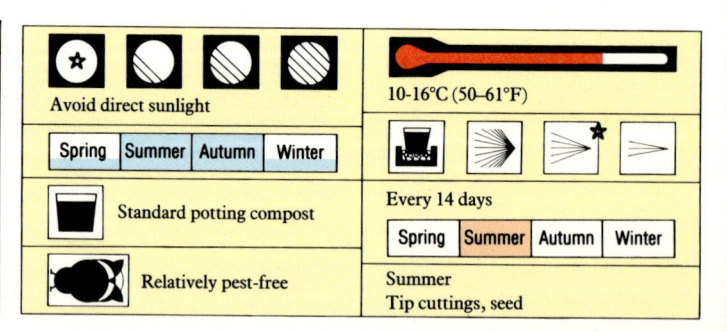

Colour Types
  Predominant single colour
    Blue
    Orange
    Pink
    Purple/Mauve
    Red
    White ✱
    Yellow
  Multicoloured

Avoid direct sunlight

| Spring | Summer | Autumn | Winter |

Standard potting compost

Relatively pest-free

10-16°C (50-61°F)

Every 14 days

| Spring | Summer | Autumn | Winter |

Summer
Tip cuttings, seed

Chrysanthemum

Japan/China

Florist's Mum

Compositae

A beautiful flowering plant which comes in a variety of colours and flower shapes. Chrysanthemums do well in a bright, cool position out of the sun. During flowering the pot should be kept moist. When planted in the garden a higher degree of excellent flowers may be obtained by using a low-nitrogen fertilizer two or three times during the growing season. After flowering, the plants can be thrown away or placed in the garden, where they will grow larger before flowering again in autumn. Plants not hardy enough to survive the winter can be brought indoors for flowering in late autumn.

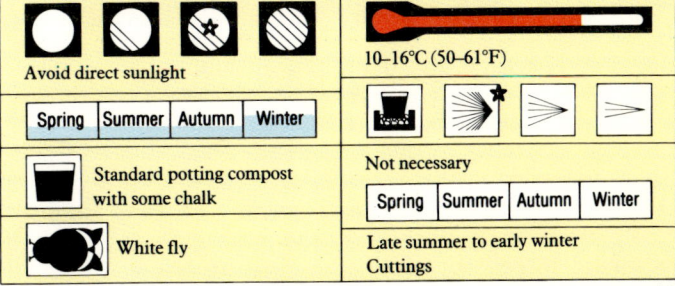

Colour Types
  Predominant single colour
    Blue
    Orange ✸
    Pink
    Purple/Mauve
    Red ✸
    White ✸
    Yellow ✸
  Multicoloured

Avoid direct sunlight

| Spring | Summer | Autumn | Winter |

Standard potting compost with some chalk

White fly

10–16°C (50–61°F)

Not necessary

| Spring | Summer | Autumn | Winter |

Late summer to early winter
Cuttings

43

Cineraria (Senecio cruentus)

Canary Islands

Bride's posy

Compositae

An annual plant, so after flowering it should really be discarded. The flowers are mauve with white centres and the anthers (pollen-bearing parts) are quite noticeable. Red, blue, white and pink flowering varieties are also available. The leaves are pale green and maple-like in shape. Ensure cool conditions as they tend to dry out easily, and place in a shady position, not in full sun. Allow the soil to remain evenly moist at all times using tepid water. A combination of these will produce a longer flowering time. Avoid draughty and warm surroundings as this can lead to an infestation of aphids. Feed only whilst flowering and during summer.

Cineraria (Senecio cruentus)

Canary Islands

Bride's posy

Compositae

C. Hurst's 'Monarch' strain

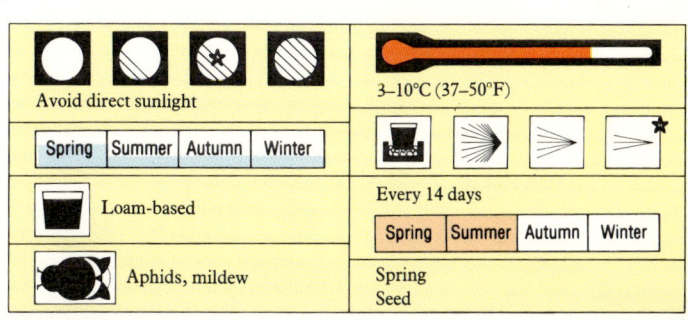

Colour Types
    Predominant single colour
        Blue ✿
        Orange
        Pink ✿
        Purple/Mauve ✿
        Red ✿
        White ✿
        Yellow
    Multicoloured

Avoid direct sunlight

| Spring | Summer | Autumn | Winter |

Loam-based

Aphids, mildew

3–10°C (37–50°F)

Every 14 days

| Spring | Summer | Autumn | Winter |

Spring
Seed

Citrus mitis

Orange tree

China

Rutaceae

Privet-like dark green leaves and small white flowers which develop into fruits. The fruits take about a year before they turn completely orange. Place in good light and air. During summer, water liberally, but keep the soil just moist in winter. Only feed during summer on a regular basis, and maintain a moderate humidity. Can be placed outdoors in the warm months. A low temperature in winter will encourage blooms. Tip cuttings taken in spring will root with bottom heat.

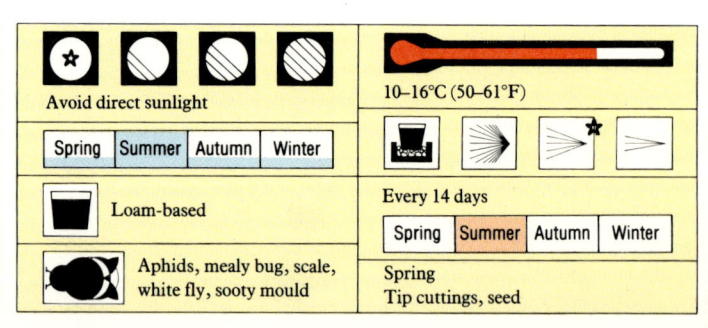

Colour Types
Predominant single colour
Blue
Orange
Pink
Purple/Mauve
Red
White ✱
Yellow
Multicoloured

Avoid direct sunlight

| Spring | Summer | Autumn | Winter |

Loam-based

Aphids, mealy bug, scale, white fly, sooty mould

10–16°C (50–61°F)

Every 14 days

| Spring | Summer | Autumn | Winter |

Spring
Tip cuttings, seed

Clerodendrum thomsoniae

Africa

Bleeding heart vine

Verbenaceae

A climbing plant with large cordate (heart-shaped) leaves with both red and white flowers, the latter being the predominating colour. Grow in a bright spot out of direct sunlight. The soil should be allowed to dry between waterings in winter, but keep well-watered the rest of the year. Feed during the summer only. As a high humidity is necessary, the foliage should be sprayed regularly. Pruning and pinching-out the tips will produce bushy, compact plants. Once it has flowered, cut back and repot. Best grown in greenhouses.

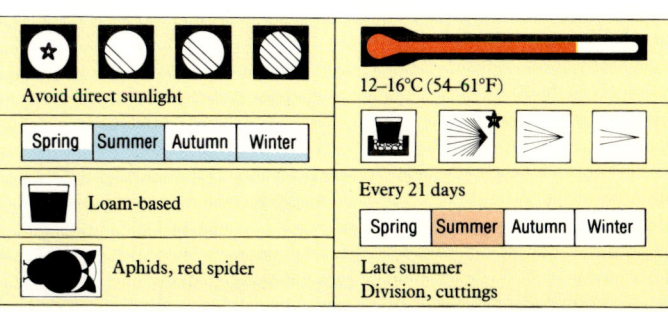

Colour Types
  Predominant single colour
    Blue
    Orange
    Pink
    Purple/Mauve
    Red ✿
    White ✿
    Yellow
  Multicoloured

Avoid direct sunlight

| Spring | Summer | Autumn | Winter |

Loam-based

Aphids, red spider

12–16°C (54–61°F)

Every 21 days

| Spring | Summer | Autumn | Winter |

Late summer
Division, cuttings

Crocus chrysanthus

Crocus

Europe

Iridaceae

A popular houseplant whose bell-shaped, upright flowers come in a variety of colours. The bulbs are potted in late autumn and kept in total darkness until the leaves form. Then reposition in a well-lit spot out of direct sunlight. Feed when the leaves have fully developed and when the buds are produced. Keep the plants fairly cool and mist from time to time to keep a high degree of humidity. When the flowers have died, store the plant for two months to allow a resting period and repot.

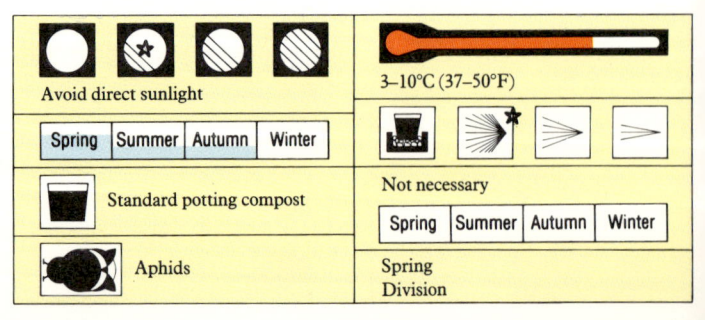

Colour Types
- Predominant single colour
  - Blue
  - Orange
  - Pink
  - Purple/Mauve ✿
  - Red
  - White ✿
  - Yellow ✿
- Multicoloured

Avoid direct sunlight

| Spring | Summer | Autumn | Winter |

Standard potting compost

Aphids

3–10°C (37–50°F)

Not necessary

| Spring | Summer | Autumn | Winter |

Spring Division

Crossandra undulifolia

Firecracker flower

India

Acanthaceae

A compact plant with large, glossy green, spear-shaped leaves, whose flowering portion (or inflorescence) develops a skirt of orange petals. Requires moderate shade in summer, but a well-lit position, out of direct sunlight, is essential during the winter. Keep the soil moist during summer and water less in winter. Feed fortnightly. A high humidity is essential if they are to thrive. Grow in combination with other plants in the same container for best results. Seeds should be planted in spring, cuttings will root in bottom heat. Ensure plants are well-ventilated.

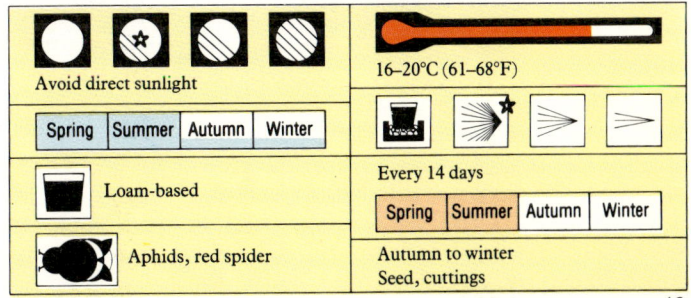

Colour Types
Predominant single colour
Blue
Orange ✱
Pink
Purple/Mauve
Red
White
Yellow
Multicoloured

Avoid direct sunlight

| Spring | Summer | Autumn | Winter |

Loam-based

Aphids, red spider

16–20°C (61–68°F)

Every 14 days

| Spring | Summer | Autumn | Winter |

Autumn to winter
Seed, cuttings

49

Cyclamen

Alpine violet

Mediterranean

Primulaceae

A popular winter pot plant whose splayed butterfly-like flowers are offset by the dark green, heart-shaped leaves. During the growing season the plant needs to be kept cool and the soil moist. Dunk in a fertilizer solution weekly. The plant grows from a tuberous root which must be placed slightly above the surface of the soil or water may become trapped in the crown and rot will follow. After flowering, decrease watering until all the leaves wilt, rest for approximately four weeks, then repot after discarding the dead foliage and gradually nurture into fresh growth.

Cyclamen

Alpine violet

Mediterranean

Primulaceae

Scented silver-leaf strain

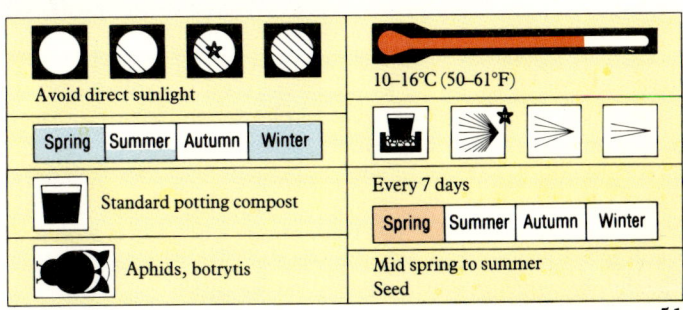

Colour Types		
Predominant single colour		
Blue		
Orange		
Pink ✱		
Purple/Mauve ✱		
Red ✱		
White ✱		
Yellow		
Multicoloured		

Avoid direct sunlight

| Spring | Summer | Autumn | Winter |

Standard potting compost

Aphids, botrytis

10–16°C (50–61°F)

Every 7 days

| Spring | Summer | Autumn | Winter |

Mid spring to summer
Seed

51

Euphorbia pulcherrima

Poinsettia

Mexico

Euphorbiaceae

This plant has large, leafy bracts available in fiery red, white or pink. The flowers themselves are small and insignificant. The plant needs to be in a well-lit position, but not in direct sunlight. During winter, when in bloom, the soil must be kept moist and the plant fed occasionally. Over-watering will result in the leaves falling after turning yellow. Flowers are encouraged by not allowing any artificial light to reach the plant or more leaves will follow. In spring, the plant, in its pot, may be placed in the garden (after being cut back). In autumn, it should be placed in complete darkness for fourteen hours daily for forty days to encourage blooms. The milky juice which results from damaged leaves or stems is poisonous, so handle with care.

Euphorbia pulcherrima — Mexico

Poinsettia — Euphorbiaceae

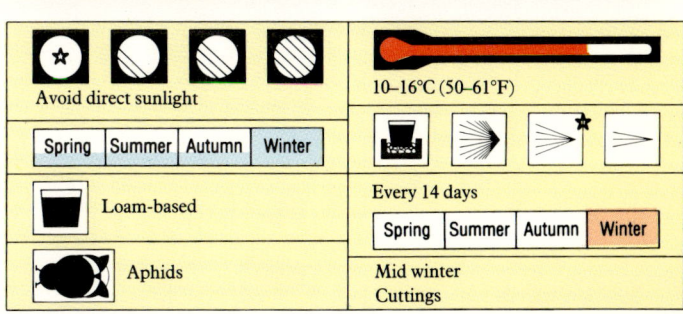

Colour Types
Predominant single colour
- Blue
- Orange
- Pink ✿
- Purple/Mauve
- Red ✿
- White ✿
- Yellow

Multicoloured

Avoid direct sunlight

| Spring | Summer | Autumn | Winter |

Loam-based

Aphids

10–16°C (50–61°F)

Every 14 days

| Spring | Summer | Autumn | Winter |

Mid winter Cuttings

53

Fuchsia

Central America

Lady's eardrop

Onagraceae

A popular bushy plant whose bell-like flowers come in a variety of colours and mixtures. The plant is best seen in hanging baskets which complement its natural growth habits. Fuchsias grow well in bright light, and should be fed. Flowers are encouraged by spray-misting regularly, giving adequate, good ventilation and supplying cool conditions. A fall in the number of buds produced is the result of poor light and high temperatures. Can be planted in the garden once winter is over, although it may take slightly longer to flower. Check for pests as they are susceptible. Cuttings can be taken throughout the summer. Difficult plants to nurture but their graceful display is extremely rewarding.

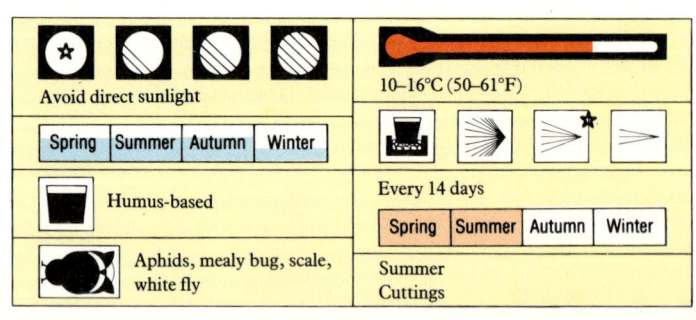

Colour Types
- Predominant single colour
  - Blue
  - Orange
  - Pink ✲
  - Purple/Mauve ✲
  - Red ✲
  - White ✲
  - Yellow
- Multicoloured

Avoid direct sunlight

| Spring | Summer | Autumn | Winter |

Humus-based

Aphids, mealy bug, scale, white fly

10–16°C (50–61°F)

Every 14 days

| Spring | Summer | Autumn | Winter |

Summer Cuttings

Gardenia intermedia

Cape jasmine

China

Rubiaceae

Evergreen, compact plants with broad, spear-shaped, dark green leaves and white aromatic flowers. Best grown in a draught-free, bright window out of direct sunlight. To water, use tepid, rain or lime-free water and keep moist in summer but somewhat drier during winter. Feed half-doses during summer only. Mist the foliage regularly to maintain a humid environment. A good circulation of air together with constant temperatures will give good results. Propagation from cuttings should commence at the onset of spring.

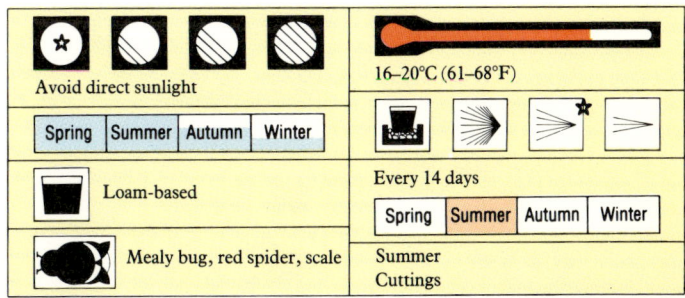

Colour Types
Predominant single colour
Blue
Orange
Pink
Purple/Mauve
Red
White ✯
Yellow
Multicoloured

Avoid direct sunlight

| Spring | Summer | Autumn | Winter |

Loam-based

Mealy bug, red spider, scale

16–20°C (61–68°F)

Every 14 days

| Spring | Summer | Autumn | Winter |

Summer
Cuttings

Gloriosa rothschildiana

Central Africa

Glory lily

Liliaceae

The striking red and yellow ruffled petals of the flowers along with the bright green leaves make this a truly beautiful example of the lily family. Position in a bright, warm window out of draughts and during the summer water liberally. Tubers should be potted in late winter/early spring. After the flowers and foliage have died, store the plants, then repot after approximately two months. When plants have reached a sufficient height, use a cane for support.

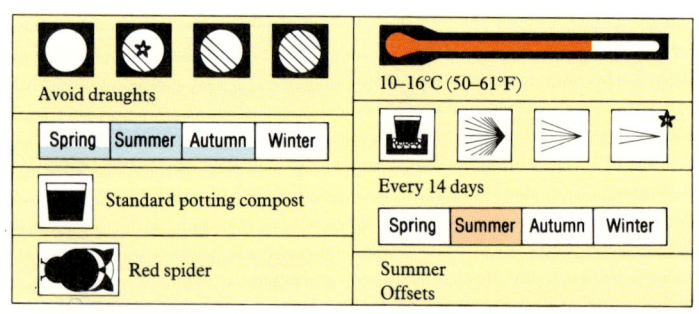

Colour Types
Predominant single colour
Blue
Orange
Pink
Purple/Mauve
Red ★
White
Yellow
Multicoloured

Avoid draughts

| Spring | Summer | Autumn | Winter |

Standard potting compost

Red spider

10–16°C (50–61°F)

Every 14 days

| Spring | Summer | Autumn | Winter |

Summer Offsets

Guzmania cardinalis

Colombia

Scarlet star

Bromeliaceae

An epiphytic bromeliad with long, green leaves and a contrasting central rosette of orange bracts with red margins. The white flowers are insignificant. Grow in good light, but do not place in direct sun. The soil should be kept moist, and the 'vase' topped up occasionally. In winter the plant will benefit if kept somewhat drier. When in flower, cool conditions will slow down the process of decay. As with most bromeliads, the offsets produced should only be potted once the parent has died off.

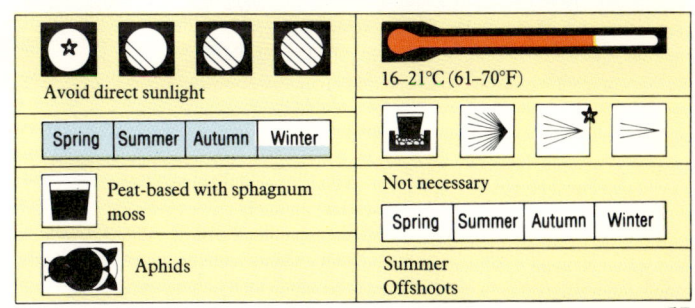

Colour Types
 Predominant single colour
  Blue
  Orange
  Pink
  Purple/Mauve
  Red ✯
  White
  Yellow
 Multicoloured

Avoid direct sunlight

| Spring | Summer | Autumn | Winter |

Peat-based with sphagnum moss

Aphids

16–21°C (61–70°F)

Not necessary

| Spring | Summer | Autumn | Winter |

Summer
Offshoots

Hibiscus

Chinese rose

China

Malvaceae

A plant that sports large, wafer-thin flowers, with a projecting cluster of golden yellow stamens. Blooms are encouraged by adequate watering and a plentiful supply of sunlight. During the growing season water well and feed fortnightly. Avoid draughts and keep a fairly even temperature or the young buds will drop off. After the flowering period, the plant should be cut back well and placed in a cool spot to rest, but do not let the soil dry out completely. Cuttings are best taken in the spring.

Hibiscus

Chinese rose

China

Malvaceae

One of the more exotic strains, *H. 'Grace God'*

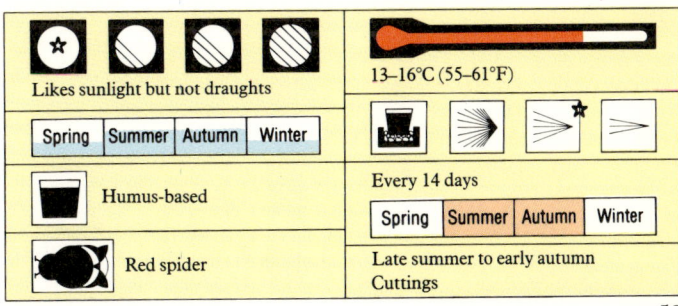

Colour Types
Predominant single colour
Blue
Orange ✻
Pink ✻
Purple/Mauve
Red ✻
White
Yellow ✻
Multicoloured

Likes sunlight but not draughts

| Spring | Summer | Autumn | Winter |

Humus-based

Red spider

13–16°C (55–61°F)

Every 14 days

| Spring | Summer | Autumn | Winter |

Late summer to early autumn
Cuttings

Hippeastrum

West Indies

Amaryllis Amaryllidaceae

A bulb plant which has large, trumpet-like flowers attached to tall, strong stems. The leaves are long and narrow, growing directly from the bulb. Flowers are available in red, pink and white. Pot the bulbs in mid winter, keeping the soil just moist and placing the pots in a warm, shady spot. When the flower stem has grown to 15-20 cm (6-8 in) tall, reposition in a well-lit spot (not in full sun) and give plenty of water. Feed only when in flower. After flowering the leaves will grow properly. Let them develop as this will ensure flowers for next season. Decrease water gradually, until the leaves die, then let the soil dry out. Sever the stem off near to the bulb. Finally, repot in mid-winter using fresh soil and start again.

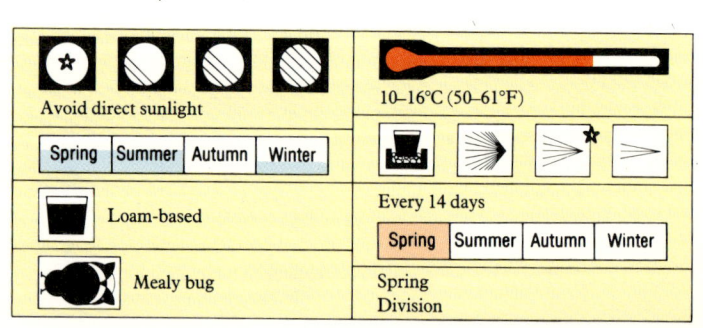

Colour Types
- Predominant single colour
    - Blue
    - Orange
    - Pink
    - Purple/Mauve
    - Red ✼
    - White ✼
    - Yellow
- Multicoloured

Avoid direct sunlight

| Spring | Summer | Autumn | Winter |

Loam-based

Mealy bug

10–16°C (50–61°F)

Every 14 days

| Spring | Summer | Autumn | Winter |

Spring
Division

Hoya bella

India

Wax flower

Asclepiadaceae

White, star-shaped, mauve-centred, aromatic flowers whose stalks all arise from the same spot, giving an umbrella-like appearance. The small leaves are a dark green. Seen at its best in hanging baskets, it requires a well-lit position out of direct sunlight. Water liberally, except in winter when the soil should be allowed to dry appreciably between waterings (also do not feed at this time). A high humidity is essential for good plants, so spray the foliage regularly. When pruning do not cut away any stems which have brought forth blooms, and avoid pruning or disturbing the plant whilst it is budding. Take cuttings in spring.

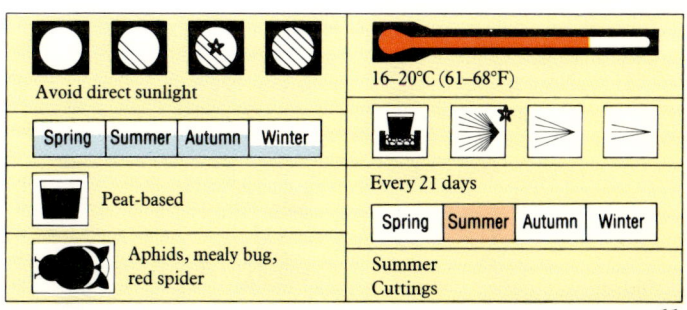

Colour Types
- Predominant single colour
    - Blue
    - Orange
    - Pink
    - Purple/Mauve
    - Red
    - White ✯
    - Yellow
- Multicoloured

Avoid direct sunlight

| Spring | Summer | Autumn | Winter |

Peat-based

Aphids, mealy bug, red spider

16–20°C (61–68°F)

Every 21 days

| Spring | Summer | Autumn | Winter |

Summer Cuttings

Hyacinthus

Asia

Hyacinth

Liliaceae

One of the most well-known and popular of bulbous plants. It has narrow, strap-like, pale green leaves surrounding a central spike which is densely 'populated' with small, bell-like, pleasantly scented sessile (stalkless) flowers. Flower colours are red, white, pink, yellow and mauve. Pot the bulbs in autumn and place in a cool, dark place for approximately two months, by which time the flower buds should be evident. Then place in a good light, but not in direct sunlight. As the plant grows, water liberally with rain or soft water. Whilst in flower, offsets will be produced. These should be allowed to develop as the original bulb will not flower again.

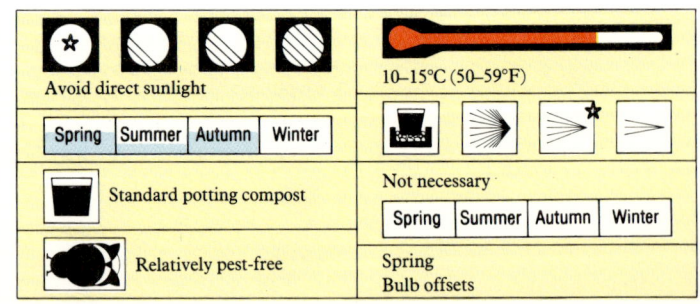

Colour Types
   Predominant single colour
      Blue ✱
      Orange
      Pink ✱
      Purple/Mauve ✱
      Red ✱
      White ✱
      Yellow ✱
   Multicoloured

Avoid direct sunlight

| Spring | Summer | Autumn | Winter |

Standard potting compost

Relatively pest-free

10–15°C (50–59°F)

Not necessary

| Spring | Summer | Autumn | Winter |

Spring
Bulb offsets

Hydrangea macrophylla — Japan

Hydrangea — Saxifragaceae

Popular plant with mop-heads of flowers in pink, red or blue. Needs good light but a cool position. Observance of soil constituents results in colour condition. Add chalk for red and aluminium sulphate for blue flowers. Cut back plants after flowering. Stem cuttings are best taken at this time. Use a hormone rooting compound to establish young plant growth. These plants require a rest period.

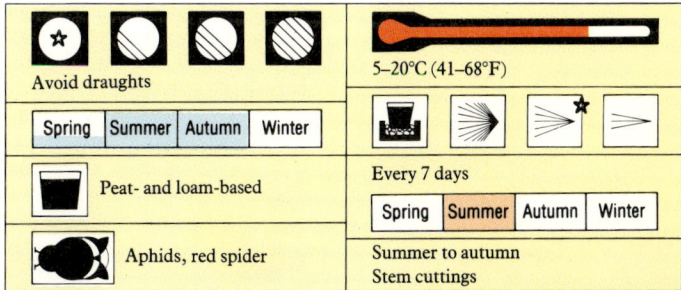

Colour Types
  Predominant single colour
    Blue ✻
    Orange
    Pink ✻
    Purple/Mauve
    Red
    White ✻
    Yellow
  Multicoloured

Avoid draughts

| Spring | Summer | Autumn | Winter |

Peat- and loam-based

Aphids, red spider

5–20°C (41–68°F)

Every 7 days

| Spring | Summer | Autumn | Winter |

Summer to autumn
Stem cuttings

Impatiens walleriana

Zanzibar

Busy Lizzie

Balsaminaceae

A fine plant with glossy, dark green leaves and red flowers. Flowers also come in white and pink. Grow in a bright position out of direct sunlight and give plenty of water during the summer, but avoid over-watering, allowing the soil to dry appreciably between waterings. Also keep on the dry side in winter. Keep temperatures fairly constant or leaf-drop will ensue. Spray occasionally to maintain humidity, but do not spray whilst in flower as stains will occur. Feed only during summer. Cuttings will readily root in water after about one month. Seeds can be sown in spring. Pinching out the growing tips will encourage bushy growth. After flowering they can be cut back drastically, at which time the top soil should be replaced with new compost.

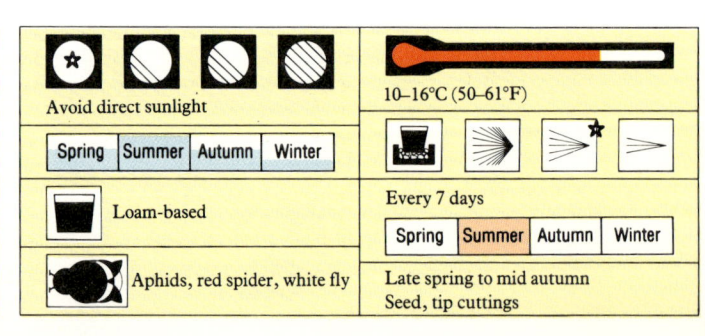

Colour Types
　Predominant single colour
　　Blue
　　Orange
　　Pink ✿
　　Purple/Mauve
　　Red ✿
　　White ✿
　　Yellow
　Multicoloured

Avoid direct sunlight

| Spring | Summer | Autumn | Winter |

Loam-based

Aphids, red spider, white fly

10–16°C (50–61°F)

Every 7 days

| Spring | Summer | Autumn | Winter |

Late spring to mid autumn
Seed, tip cuttings

Jasminum

China

Jasmine

Oleaceae

A lovely aromatic plant with small, green sessile leaves with a profusion of star-shaped, white flowers. Yellow varieties are also available. Good sun is the order of the day when growing jasmines or they will not flower. An airy position is also necessary. The pots should be given a good dunking in water at least once a month. Monthly feeding whilst the plants are growing is advisable. Regular spraying of the foliage is also helpful, as this ensures a high humidity which is essential for healthy plants. After the flowers have died, prune well back. Take cuttings in spring. Plants can be placed outdoors in summer.

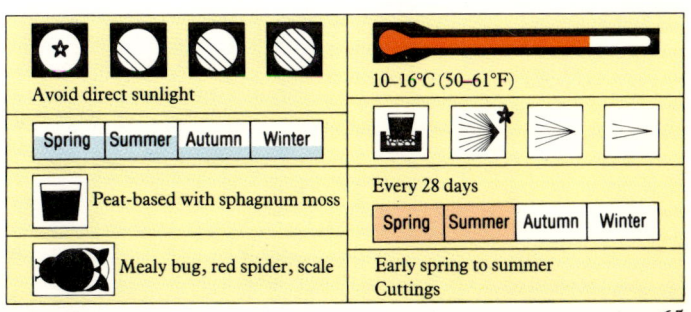

Colour Types
　Predominant single colour
　　Blue
　　Orange
　　Pink
　　Purple/Mauve
　　Red
　　White ✿
　　Yellow ✿
　Multicoloured

Avoid direct sunlight

| Spring | Summer | Autumn | Winter |

Peat-based with sphagnum moss

Mealy bug, red spider, scale

10–16°C (50–61°F)

Every 28 days

| Spring | Summer | Autumn | Winter |

Early spring to summer
Cuttings

65

Kalanchoe blossfeldiana

Tom Thumb

Asia

Crassulaceae

A plant with leathery leaves with a flatish head of brilliant scarlet flowers on a green stem. The beauty of this plant is that flowers can appear twice in a year and for long spells. This succulent grows well in a well-lit spot, but avoid direct sunlight. Water regularly when the plants are in bloom but be sure not to wet the leaves, as this will cause them to rot very quickly. After the first flowering, restrict the light to eight hours a day and blooms should appear again in three to four months.

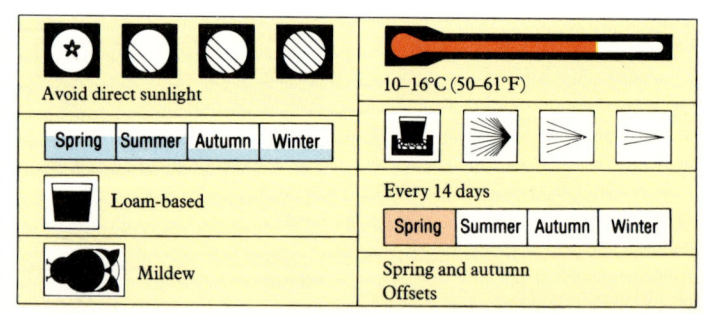

Colour Types
    Predominant single colour
        Blue
        Orange
        Pink
        Purple/Mauve
        Red ✲
        White
        Yellow
    Multicoloured

Avoid direct sunlight

| Spring | Summer | Autumn | Winter |

Loam-based

Mildew

10–16°C (50–61°F)

Every 14 days

| Spring | Summer | Autumn | Winter |

Spring and autumn
Offsets

Lilium 'Enchantment'   Japan

Lily   Liliaceae

Bulbous plants which produce tall, erect stems topped by stunning flowers in a variety of colour patterns in red and white. For spring flowers, autumn is the best time for potting. Once growth has reached about 10 cm (4 in) place in good light. Whilst in growth, the soil should be kept moist until they have flowered, then water sparingly. Repot annually in fresh soil. Keep conditions cool and avoid placing near heat sources. Ensure that the plants receive good ventilation.

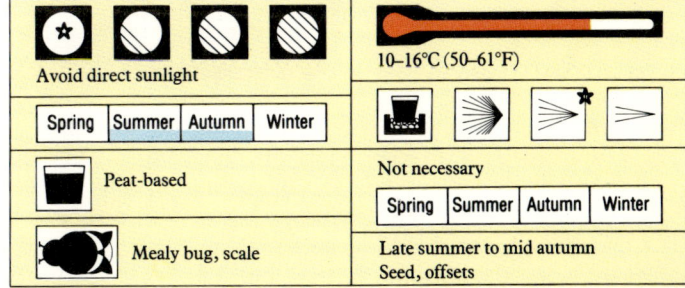

Colour Types
Predominant single colour
Blue
Orange ✿
Pink
Purple/Mauve
Red ✿
White
Yellow
Multicoloured

Avoid direct sunlight

Spring	Summer	Autumn	Winter

Peat-based

Mealy bug, scale

10–16°C (50–61°F)

Not necessary

Spring	Summer	Autumn	Winter

Late summer to mid autumn
Seed, offsets

Narcissus

Asia/Europe

Daffodil   Amaryllidaceae

Beautiful plants whose narrow, strap-like leaves and delicate flowers make them a decorative addition to any household. Pot the bulbs in pebbles or compost in shallow bowls. The tips of the bulbs should be no less than 3 cm (1 in) below the topsoil. Place the bulbs in a dark spot for about two months, when the leaves should be about 10 cm (4 in) high. Then reposition in good light and warmth. Ensure that there is ample water when the plants are growing.

Narcissus

Daffodil

Asia/Europe

Amaryllidaceae

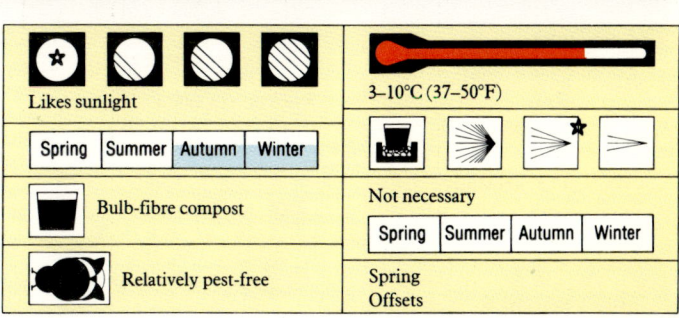

Colour Types
Predominant single colour
Blue
Orange ✿
Pink
Purple/Mauve
Red
White ✿
Yellow ✿
Multicoloured ✿

Likes sunlight

| Spring | Summer | Autumn | Winter |

Bulb-fibre compost

Relatively pest-free

3–10°C (37–50°F)

Not necessary

| Spring | Summer | Autumn | Winter |

Spring
Offsets

69

Orchidaceae

South America/Asia

Orchid

Orchidaceae

There are well over 30,000 species of orchid and the ease at which most varieties interbreed has resulted in some 60,000 artificial hybrids. The flowerhead consists of a ring of three petals and three sepals. The middle petal is called the 'labellum', or lip, and is usually differently coloured and marked than the other two petals. Given the right conditions and care, the results are very rewarding. Keep in a good light, but avoid direct sunlight. A drop in temperature at night is essential for good flowering. Maintain a high humidity level – create a micro-climate. Watering requirements vary for the different species: Cattleya, Oncidium and Miltonia should approach dryness prior to watering, whilst others like Cymbidium, Paphiopedilum and Phalaenopsis should always be kept moist. Little feeding is necessary and only very sparingly. Repot when they become overcrowded in their pots. Propagation is by division, offsets and seed.

Cattleya

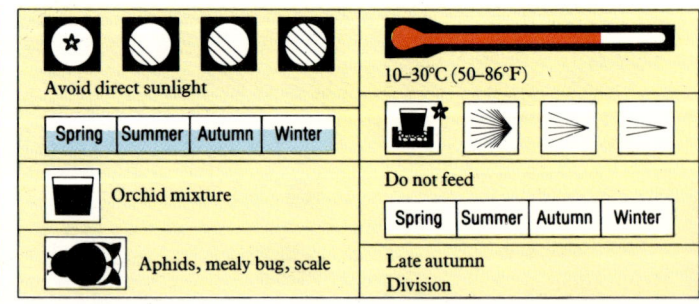

Colour Types
  Predominant single colour
    Blue
    Orange
    Pink
    Purple/Mauve
    Red
    White
    Yellow
  Multicoloured ✯

Avoid direct sunlight

| Spring | Summer | Autumn | Winter |

Orchid mixture

Aphids, mealy bug, scale

10–30°C (50–86°F)

Do not feed

| Spring | Summer | Autumn | Winter |

Late autumn
Division

70

Orchidaceae

South America/Asia

Orchid   Orchidaceae

Cymbidium

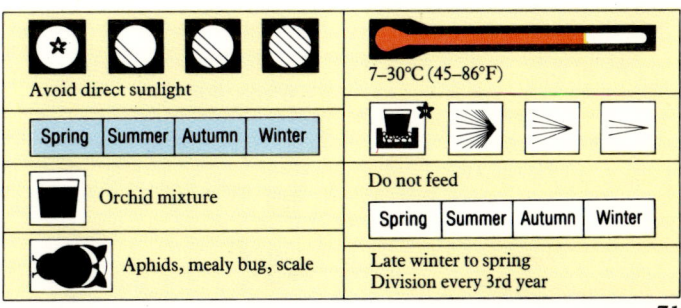

Colour Types
- Predominant single colour
  - Blue
  - Orange
  - Pink
  - Purple/Mauve
  - Red
  - White
  - Yellow
- Multicoloured ✦

Avoid direct sunlight

| Spring | Summer | Autumn | Winter |

Orchid mixture

Aphids, mealy bug, scale

7–30°C (45–86°F)

Do not feed

| Spring | Summer | Autumn | Winter |

Late winter to spring
Division every 3rd year

Orchidaceae

South America/Asia

Orchid   Orchidaceae

Miltonia

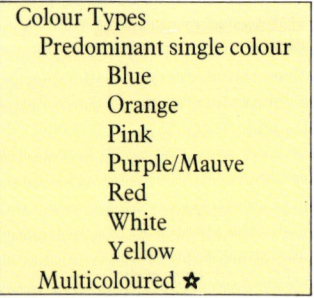

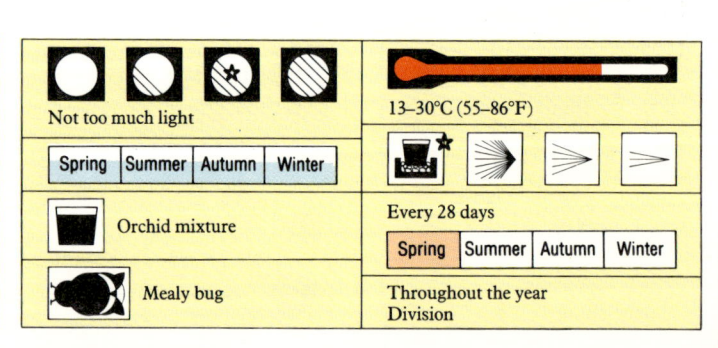

Colour Types
- Predominant single colour
  - Blue
  - Orange
  - Pink
  - Purple/Mauve
  - Red
  - White
  - Yellow
- Multicoloured ✮

Not too much light

| Spring | Summer | Autumn | Winter |

Orchid mixture

Mealy bug

13–30°C (55–86°F)

Every 28 days

| Spring | Summer | Autumn | Winter |

Throughout the year
Division

72

Orchidaceae

South America/Asia

Orchid Orchidaceae

Paphiopedilum

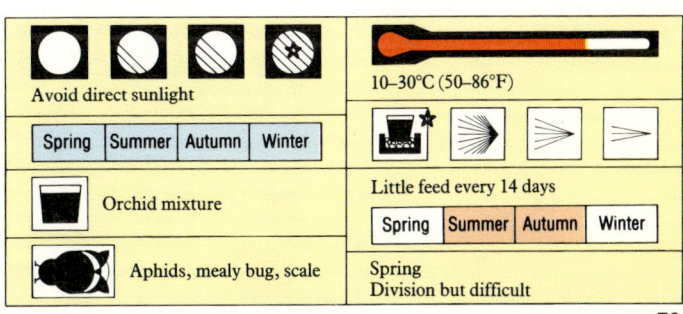

Colour Types
　Predominant single colour
　　Blue
　　Orange
　　Pink
　　Purple/Mauve
　　Red
　　White
　　Yellow
　Multicoloured ★

Avoid direct sunlight

| Spring | Summer | Autumn | Winter |

Orchid mixture

Aphids, mealy bug, scale

10–30°C (50–86°F)

Little feed every 14 days

| Spring | Summer | Autumn | Winter |

Spring
Division but difficult

73

Pachystachys lutea

Mexico

Golden shrimp plant

Acanthaceae

Somewhat similar to Aphelandra, but the leaves are unmarked and softer. The flowering part, or inflorescence, has bright yellow bracts, which last for quite a while, and white flowers. These plants require a well-lit position and good soil drainage. During summer when the plant is growing give ample water. Dunking occasionally is beneficial. Also feed monthly at this time. In winter, keep on the dry side. Cuttings, which root readily, can be taken in spring. Pruning and pinching out the growing tips will encourage bushy growth. Avoid extremes of temperature and try to maintain a humid environment.

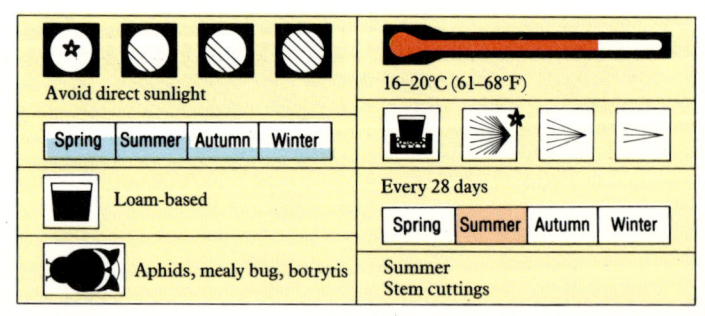

Colour Types
 Predominant single colour
  Blue
  Orange
  Pink
  Purple/Mauve
  Red
  White
  Yellow ★
 Multicoloured

Avoid direct sunlight

| Spring | Summer | Autumn | Winter |

Loam-based

Aphids, mealy bug, botrytis

16–20°C (61–68°F)

Every 28 days

| Spring | Summer | Autumn | Winter |

Summer
Stem cuttings

Passiflora caerulea — Brazil

Passion flower — Passifloraceae

An exotic vine with white, blue and mauve flowers and leaves. A warm, well-lit position is needed, as it will only flower in these conditions. After flowering, the orange-coloured fruit will become evident. Whilst in growth, the plant needs a plentiful water supply and should be fed fortnightly. But in winter, when the plant is resting, do not feed and give less water. If leaves are produced at the expense of flowers, then withhold feed immediately. It is usual for some leaves to turn yellow and drop, so long as the rest stay healthy. A suitable support is also necessary when growing these plants.

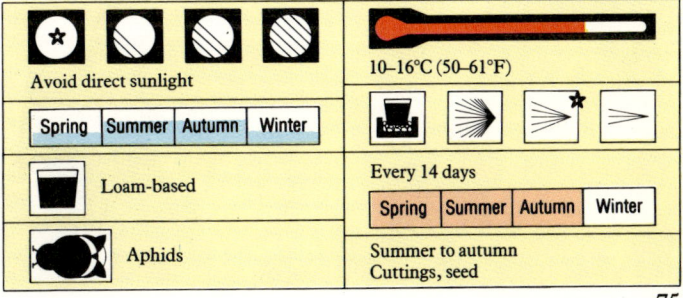

Colour Types
　Predominant single colour
　　Blue ✱
　　Orange
　　Pink
　　Purple/Mauve ✱
　　Red
　　White
　　Yellow
　Multicoloured

Avoid direct sunlight

| Spring | Summer | Autumn | Winter |

Loam-based

Aphids

10–16°C (50–61°F)

Every 14 days

| Spring | Summer | Autumn | Winter |

Summer to autumn
Cuttings, seed

Pelargonium

South Africa

Geranium

Geraniaceae

There are hundreds of species of geranium, of which three are favourites – the regals (*P. grandiflorum*), the zonals (*P. zonale*) and the ivy-leaved trailers (*P. peltatum*). The regals are shrub-like with large, floppy flowers. Colours range from white to red with flowering times around April and May. The zonals are distinguished by the dark green circular patches on their leaves. Colours vary from white to deep red with flowers from April to October. The trailers have succulent and smooth ivy-like foliage with predominant shades of red, white and pink. During the growing season, feed and keep moist. Cut back after the flowers have wilted. Terminal cuttings should be taken in mid-August.

Pelargonium

South Africa

Geranium

Geraniaceae

Colour Types
Predominant single colour
Blue
Orange
Pink ✿
Purple/Mauve ✿
Red ✿
White ✿
Yellow
Multicoloured

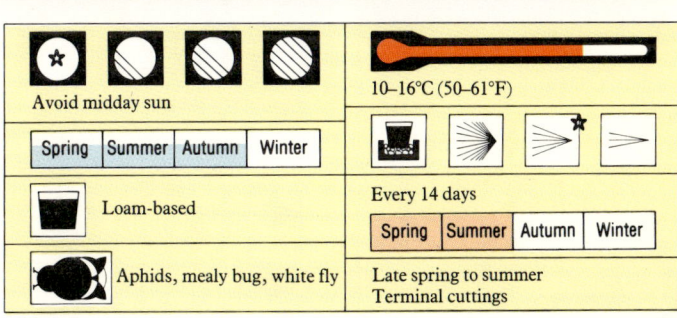

Avoid midday sun

| Spring | Summer | Autumn | Winter |

Loam-based

Aphids, mealy bug, white fly

10–16°C (50–61°F)

Every 14 days

| Spring | Summer | Autumn | Winter |

Late spring to summer
Terminal cuttings

77

Primula

Primrose

Europe

Primulaceae

A herald of spring, this plant has yellow flowers and large, broad leaves. The hybrids available produce red, pink, blue and white flowers. Place in a well-lit position, screened from full sunlight. Keep the soil well moist whilst it is growing and water more when in the flowering season. Feeding weekly at this time is advisable, as it leads to a prolonged flowering time. In winter, somewhat drier soil conditions are best. Seeds are sown in spring. Yellowing foliage means either that the mineral level in the soil is too concentrated or that the soil is too dry. In the case of the former, repotting is necessary and any subsequent water given should be slightly demineralized.

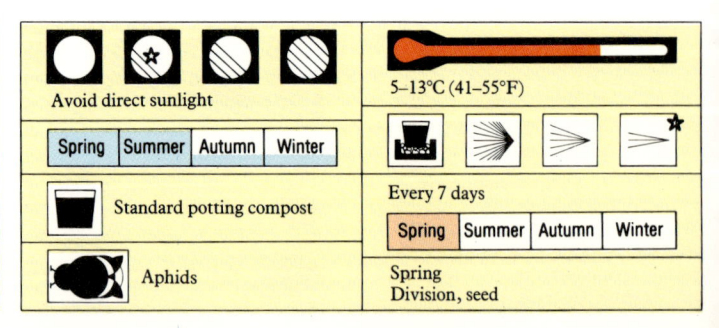

Colour Types
- Predominant single colour
    - Blue ✻
    - Orange
    - Pink ✻
    - Purple/Mauve
    - Red ✻
    - White ✻
    - Yellow ✻
- Multicoloured

Avoid direct sunlight

Spring | Summer | Autumn | Winter

Standard potting compost

Aphids

5–13°C (41–55°F)

Every 7 days

Spring | Summer | Autumn | Winter

Spring
Division, seed

Saintpaulia

African violet

East Africa

Gesneriaceae

A hairy-leaved perennial whose range of coloured flowers makes for a very popular indoor plant. It flowers abundantly at any time of the year. Feed once a month except in winter. Avoid splashing water on the leaves, as they will stain easily. Give less water and allow to rest after flowering, as this will encourage new buds to be produced. Keep humidity high as dry air causes bud-drop and curling leaves. Turn the pots regularly to ensure an equal distribution of light. Avoid direct sunlight. Cuttings are best taken in spring.

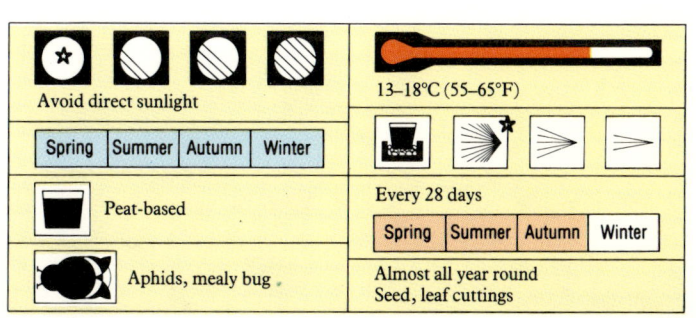

Colour Types
- Predominant single colour
  - Blue
  - Orange
  - Pink ★
  - Purple/Mauve ★
  - Red ★
  - White ★
  - Yellow
  - Multicoloured ★

Avoid direct sunlight

| Spring | Summer | Autumn | Winter |

Peat-based

Aphids, mealy bug

13–18°C (55–65°F)

Every 28 days

| Spring | Summer | Autumn | Winter |

Almost all year round
Seed, leaf cuttings

Sinningia speciosa

Brazil

Gloxinia

Gesneriaceae

Large bell-shaped flowers in white, red, pink and mauve, and broad, hairy oblong leaves which are sometimes red-tinged on the underside are the distinguishing features of *Sinningia speciosa*. Pot the tubers in autumn or spring, hollow side up, with the soil barely covering them and maintain an even moisture. Once the tubers start growing, move them to a warm spot and give more water. Grow *S. speciosa* in a well-lit position, out of direct sunlight. Use tepid water, and feed in summer only when the buds have formed. After the flowers have died, cut back any leaves and allow the soil to dry gradually, then store the tubers. Let them rest for about two months, then finally repot in fresh soil. Always ensure adequate humidity and do not over-water as this can kill the plants.

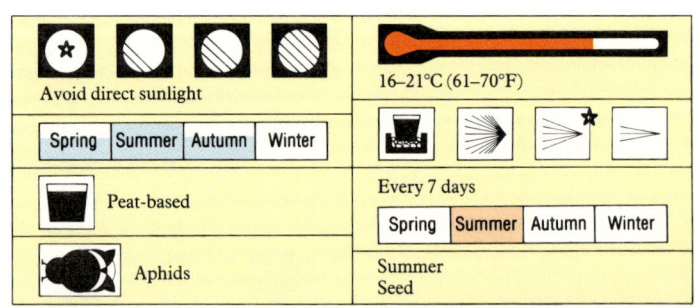

Colour Types
Predominant single colour
Blue
Orange
Pink ✱
Purple/Mauve ✱
Red ✱
White ✱
Yellow
Multicoloured

Avoid direct sunlight

| Spring | Summer | Autumn | Winter |

Peat-based

Aphids

16–21°C (61–70°F)

Every 7 days

| Spring | Summer | Autumn | Winter |

Summer
Seed

Solanum capsicastrum **Brazil**

Christmas cherry Solanceae

Small, white, star-shaped flowers (followed ultimately by the shiny red fruits) with dark leaves aptly describes this cheerful shrub. When the flowers have formed (only one in each umbrella-shaped canopy is fertile), pollination by hand should be carried out to achieve better fruiting results. Cool conditions will prolong the life of the fruits. When in flower, water liberally and feed regularly in summer, but during winter keep the plants on the dry side. When the leaves fall, the plant should be pruned and rested. In early spring growth will begin anew and the plants can be put outside throughout the summer and then brought back inside before the frosty weather sets in.

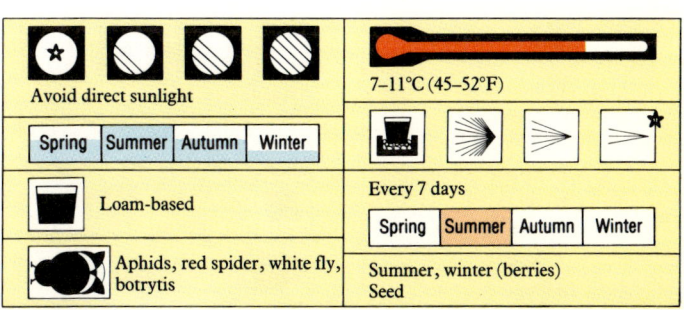

Colour Types
  Predominant single colour
    Blue
    Orange
    Pink
    Purple/Mauve
    Red
    White ✱
    Yellow
  Multicoloured

Avoid direct sunlight

| Spring | Summer | Autumn | Winter |

Loam-based

Aphids, red spider, white fly, botrytis

7–11°C (45–52°F)

Every 7 days

| Spring | Summer | Autumn | Winter |

Summer, winter (berries)
Seed

Spathiphyllum wallisii

Colombia

Peace lily

Araceae

An exquisite member of the lily family with a white ovate flower and a bright yellow spike (spadix) bearing both male and female flowers. The leaves are spear-shaped with ruffled edges and are bright green. *Spathiphyllum wallisii* is best grown in semi-shade during the summer and in a bright location in winter. Ensure that the soil drains freely, and allow it to dry a little between waterings. Maintain a high humidity (regular spraying helps, this will also discourage red spider mite). Plants are repotted in early spring, and propagation is best carried out at this time. Can be placed outdoors temporarily whilst it is in flower.

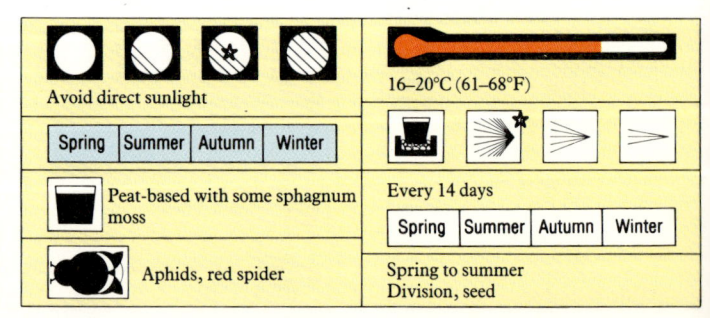

Colour Types
- Predominant single colour
  - Blue
  - Orange
  - Pink
  - Purple/Mauve
  - Red
  - White ✶
  - Yellow
- Multicoloured

Avoid direct sunlight

| Spring | Summer | Autumn | Winter |

Peat-based with some sphagnum moss

Aphids, red spider

16–20°C (61–68°F)

Every 14 days

| Spring | Summer | Autumn | Winter |

Spring to summer
Division, seed

Strelitzia reginae

Bird of paradise flower

South Africa

Musaceae

A beautiful exotic plant having flat-bladed leaves with a flower shaped like a green-hued canoe from which a crest of orange and blue petals protrude, resembling the crest of the bird of paradise. Keep in a sunny position and grow in large pots (this is crucial for the development of blooms). The soil should be kept moist in the growing period, but water very moderately in the winter.

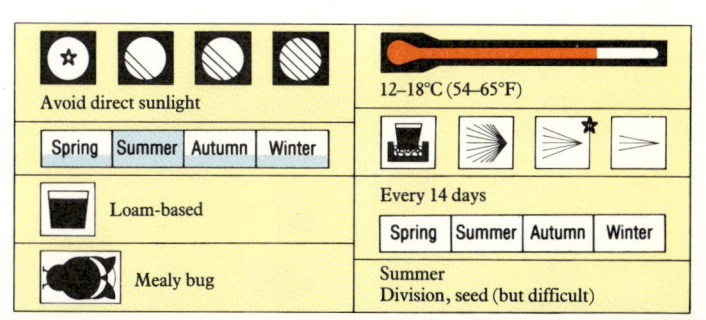

Colour Types
- Predominant single colour
  - Blue
  - Orange ✶
  - Pink
  - Purple/Mauve
  - Red
  - White
  - Yellow
- Multicoloured

Avoid direct sunlight

| Spring | Summer | Autumn | Winter |

Loam-based

Mealy bug

12–18°C (54–65°F)

Every 14 days

| Spring | Summer | Autumn | Winter |

Summer
Division, seed (but difficult)

83

Torenia fournieri

Wishbone flower

South America

Scrofulariaceae

Torenia fournieri has pale green, lanceolate leaves and single flowers borne at the end of branches or stems. The flowers are violet with purple blotches and are trumpet-like in appearance. Can be placed in moderate shade or good sun. When watering, give ample water but avoid allowing the soil to dry out. A moderate humidity is needed, and keep the temperature fairly constant. Pinching-out will encourage bushy plants. During the warm months, water liberally and feed fortnightly. Should growth appear slow, then feed once a week. Plants are best discarded once they have bloomed. Seeds are sown in late winter/early spring and will sprout with bottom heat under a thin layer of compost.

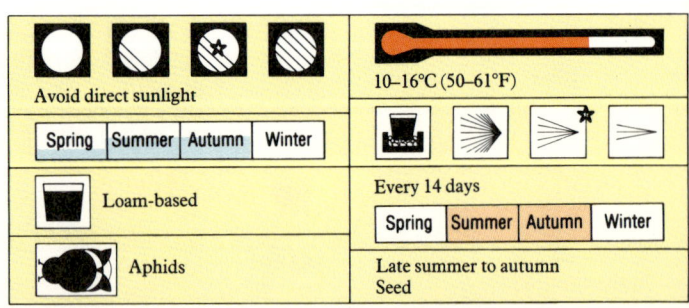

Colour Types
Predominant single colour
Blue
Orange
Pink
Purple/Mauve ✶
Red
White
Yellow
Multicoloured

Avoid direct sunlight

| Spring | Summer | Autumn | Winter |

Loam-based

Aphids

10–16°C (50–61°F)

Every 14 days

| Spring | Summer | Autumn | Winter |

Late summer to autumn
Seed

Tulipa

Asia/Europe

Tulip

Liliaceae

A popular houseplant which is available in a variety of colours. These bulbous, cup-like flowers have sessile, pale green leaves on upright stems. Pot the bulbs in autumn in moist compost and place in a cool, shaded location. Once the flower bud has developed, reposition in good light and warmth. Water frugally, keeping the soil just moist (misting the bud is also beneficial). Keep temperature constant, as if it gets too high the buds will die. Propagation is from offsets taken once the flowers have faded.

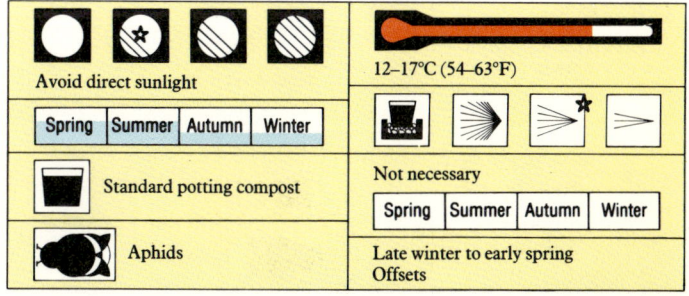

Colour Types
- Predominant single colour
    - Blue
    - Orange ✿
    - Pink ✿
    - Purple/Mauve
    - Red ✿
    - White ✿
    - Yellow ✿
- Multicoloured ✿

Avoid direct sunlight

| Spring | Summer | Autumn | Winter |

Standard potting compost

Aphids

12–17°C (54–63°F)

Not necessary

| Spring | Summer | Autumn | Winter |

Late winter to early spring
Offsets

Vriesia splendens

Guyana

Flaming sword

Bromeliaceae

A fine bromeliad whose red-bracted, sword-like inflorescence rises above the rosette of broad, brown-banded, spear-shaped leaves. The flowers are yellow and are evident in summer. When growing *Vriesia splendens* ensure adequate light and keep the temperature fairly constant. The plant should also be misted regularly as this will help to maintain humidity. The offsets, at the base of the rosette, appear after the plant has flowered, and they may be potted separately once they have grown to about 10-15 cm (4-6 in) in height. Check for red spider mite which will be encouraged by hot, dry conditions.

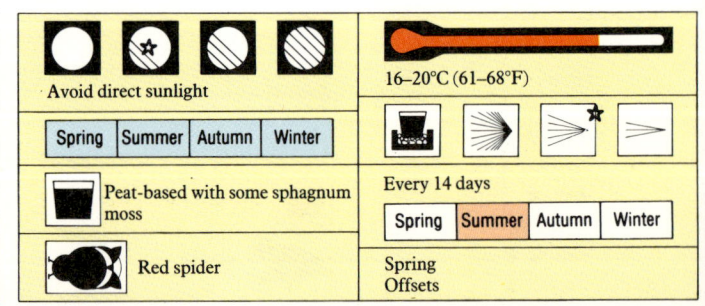

Colour Types
- Predominant single colour
  - Blue
  - Orange
  - Pink
  - Purple/Mauve
  - Red ✱
  - White
  - Yellow
- Multicoloured

Avoid direct sunlight

| Spring | Summer | Autumn | Winter |

Peat-based with some sphagnum moss

Red spider

16–20°C (61–68°F)

Every 14 days

| Spring | Summer | Autumn | Winter |

Spring Offsets

Zantedeschia aethiopica

South Africa

Arum lily

Araceae

A fine tuberous lily with dark green, spear-shaped leaves and large, white flower spathes. Let the plant die off after flowering and repot in fresh compost before storing in a cool position.

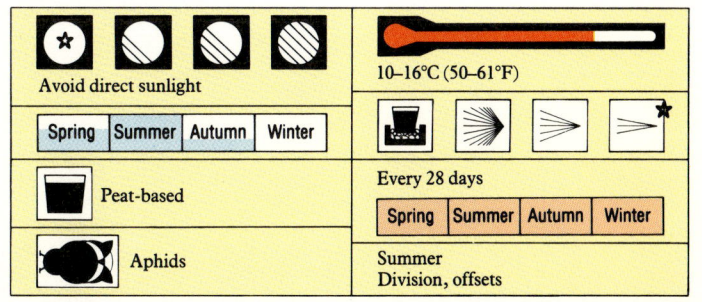

Colour Types
- Predominant single colour
    - Blue
    - Orange
    - Pink
    - Purple/Mauve
    - Red
    - White ✱
    - Yellow
- Multicoloured

Avoid direct sunlight

| Spring | Summer | Autumn | Winter |

Peat-based

Aphids

10–16°C (50–61°F)

Every 28 days

| Spring | Summer | Autumn | Winter |

Summer
Division, offsets

87

Zygocactus truncatus

Brazil

Crab cactus

Cactaceae

Often referred to as Schlumbegera, the flattened, pendant, jointed leaf stems with saw-toothed margins give the plant a crab-like appearance. Flowers are formed at the tips of the branches, and come in single shades of either white, red, orange or pink. The flowering period is triggered off by the amount of light received, and in natural conditions these plants will blossom in late autumn. Best grown in a well-lit position out of direct sunlight. Keep soil fairly dry until buds appear, then increase watering. Do not reposition plants. Take stem cuttings after blooms have fallen. These should be placed in sphagnum moss and water withheld until signs of wilting or roots appear, then treat as a mature plant.

Zygocactus truncatus   **Brazil**

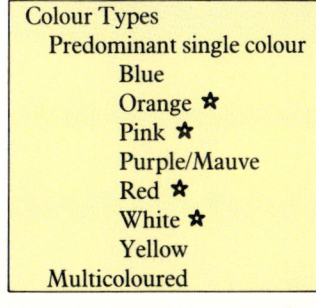

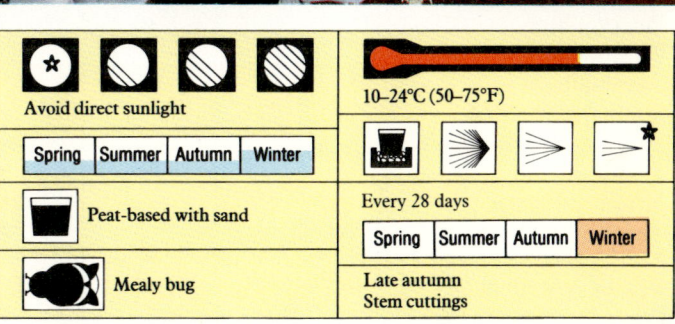

Crab cactus   Cactaceae

Colour Types
Predominant single colour
- Blue
- Orange ✪
- Pink ✪
- Purple/Mauve
- Red ✪
- White ✪
- Yellow

Multicoloured

Avoid direct sunlight

| Spring | Summer | Autumn | Winter |

Peat-based with sand

Mealy bug

10–24°C (50–75°F)

Every 28 days

| Spring | Summer | Autumn | Winter |

Late autumn
Stem cuttings

Table of flowering seasons and colour types

Flowering season

	Spring	Summer	Autumn	Winter
Acalypha hispida	★	★		
Aechmea fasciata	★			
Aeschynanthus	★			
Allamanda cathartica	★			
Anthurium scherzerianum	★	★		
Aphelandra squarrosa 'Dania'			★	
Azalea	★			★
Begonia		★	★	
Beloperone guttata	★	★	★	
Bougainvillea	★			
Cactaceae		Various		
Calceolaria	★	★		
Camellia japonica	★			★
Canna indica		★	★	
Catharanthus roseus		★		
Chrysanthemum		★	★	
Cineraria	★			
Citrus mitis	★			
Clerodendrum thomsoniae			★	
Crocus chrysanthus	★			
Crossandra undulifolia			★	★
Cyclamen	★			★
Euphorbia pulcherrima				★
Fuchsia		★		
Gardenia intermedia		★		
Gloriosa rothschildiana		★		
Guzmania cardinalis		★		
Hibiscus		★	★	
Hippeastrum	★			
Hoya bella		★		
Hyacinthus	★			
Hydrangea macrophylla		★	★	
Impatiens walleriana		★	★	
Jasminum	★	★		
Kalanchoe blossfeldiana	★		★	
Lilium 'Enchantment'		★	★	
Narcissus	★			
Orchidaceae		Various		
Pachystachys lutea		★		
Passiflora caerulea		★	★	
Pelargonium	★	★		
Primula	★			
Saintpaulia		Nearly all year		
Sinningia speciosa		★		
Solanum capsicastrum		★		
Spathiphyllum wallisii	★	★		
Strelitzia reginae		★		
Torenia fournieri		★	★	
Tulipa	★			★
Vriesia splendens	★			
Zantedeschia aethiopica		★		
Zygocactus truncatus			★	★

Predominant single colour								Multicoloured
Blue	Orange	Pink	Purple/Mauve	Red	White	Yellow	
				★			
				★			
				★		★	
		★				★	
		★	★	★	★	★	Also in white
	★	★		★	★	★	
		★					
			★				
	★	★	★	★	★	★	
	★				★	★	
		★		★	★		
					★		
					★		
	★			★	★	★	
★		★	★	★	★		
					★		
	★						
		★	★	★	★	★	
		★		★	★		
		★	★	★	★		Also combinations
					★		
				★			
				★			
	★	★		★		★	
				★	★		
					★		
★		★	★	★	★	★	
★		★			★		
		★		★	★		
					★	★	
					★	★	
	★						
	★				★	★	Also combinations
		★	★	★	★	★	Also combinations
						★	
★							
★		★	★	★	★		
		★		★	★	★	Also combinations
		★	★	★	★		
		★	★	★	★		
					★		
					★		
	★						
			★				
	★	★		★	★	★	Also combinations
				★			
					★		
	★	★		★	★		

Index of common names

African violet	79	*Hydrangea*	63
Alpine violet	50	*Japanese rose*	29
Amaryllis	60	*Jasmine*	65
Arum lily	87	*Lady's eardrop*	54
Basket vine	24	*Lady's slipper*	38
Bird of paradise flower	83	*Lily*	67
Bleeding heart vine	47	*Lipstick vine*	25
Bride's posy	44	*Madagascar periwinkle*	42
Busy Lizzie	64	*Orange tree*	46
Cactus	34	*Orchid*	70-73
Canna	41	*Paper flower*	33
Cape jasmine	55	*Paper petals*	30
Chinese rose	58	*Passion flower*	75
Christmas cherry	81	*Peace lily*	82
Crab cactus	88	*Poinsettia*	52
Crocus	48	*Primrose*	78
Daffodil	68	*Red hot cat's tail*	22
Firecracker flower	49	*Scarlet star*	57
Flaming sword	86	*Shrimp plant*	32
Flamingo plant	27	*Tea plant*	40
Florist's Mum	43	*Tom Thumb*	66
Geranium	76	*Tulip*	85
Glory lily	56	*Urn plant*	23
Gloxinia	80	*Wax flower*	61
Golden shrimp plant	74	*Wishbone flower*	84
Golden trumpet	26	*Zebra plant*	28
Hyacinth	62		

The Publishers would like to thank the following individuals and organizations for their kind permission to reproduce the photographs in this book:
A-Z Collection 27, 34, 37, 43, 46, 47, 52, 53, 55, 56, 59, 61, 67, 70, 72, 75, 83, 84; Harry Smith Horticultural Photographic Collection 6/7, 23, 25, 28, 29, 30, 31, 32, 33, 35, 38/39, 45, 48, 49, 51, 57, 62, 63, 64, 65, 66, 68, 71, 74, 79, 80, 81, 82, 85, 86, 87; TMH 10/11, 54, 83c; Michael Warren Horticultural Photo Library 5, 8/9, 22, 24, 26, 36, 40, 41, 42, 44, 50, 60, 69, 73, 76, 77, 78, 88/89 and cover. Special thanks to Mark Bryant, Frances Hobbs, Jackie Strawn and Colin Swann.

All rights reserved. No part of this publication may be reproduced, stored in a retrieval system, or transmitted, in any form or by any means, electronic, mechanical, photocopying, recording, or otherwise, without prior permission of Anthony Oorloff Publications Ltd., 18 Priors Road, Windsor, Berkshire SL4 4PD.

Any correspondence concerned with the contents of this work should be addressed to Anthony Oorloff Publications Ltd., England.

Whilst every care has been taken to ascertain the effectiveness of the methods and instructions in this publication, the authors and publishers cannot guarantee any advice given, nor shall they be liable for any loss or damage in the unlikely event that it should arise from this advice.